NATURE'S PATTERNS

Philip Ball is a science writer, and the author of many popular science books including *H₂O: A Biography of Water*, *Bright Earth*, *Critical Mass* (winner of the 2005 Aventis Prize for Science Books) and *The Music Instinct*. He lectures widely and has contributed to magazines and newspapers, including *Nature*, *New Scientist*, *The Guardian*, and *The New York Times*.

NATURE'S PATTERNS

A Tapestry in Three Parts

PHILIP BALL

Nature's Patterns is a trilogy composed of
Shapes, Flow, and Branches

OXFORD
UNIVERSITY PRESS

OXFORD
UNIVERSITY PRESS

Great Clarendon Street, Oxford OX2 6DP

Oxford University Press is a department of the University of Oxford.
It furthers the University's objective of excellence in research, scholarship,
and education by publishing worldwide in

Oxford New York

Auckland Cape Town Dar es Salaam Hong Kong Karachi
Kuala Lumpur Madrid Melbourne Mexico City Nairobi
New Delhi Shanghai Taipei Toronto

With offices in

Argentina Austria Brazil Chile Czech Republic France Greece
Guatemala Hungary Italy Japan Poland Portugal Singapore
South Korea Switzerland Thailand Turkey Ukraine Vietnam

Oxford is a registered trade mark of Oxford University Press
in the UK and in certain other countries

Published in the United States
by Oxford University Press Inc., New York

British Library Cataloguing in Publication Data

Data available

Library of Congress Cataloging in Publication Data

Data available

Typeset by SPI Publisher Services, Pondicherry, India
Printed in Great Britain
on acid-free paper by
Clays Ltd., St Ives plc

ISBN 978-0-19-960487-6

1 3 5 7 9 10 8 6 4 2

FLOW

Movement creates pattern and form. Moving water arranges itself into eddies, and sometimes places these in strict array, where they become baroque and orderly conduits for unceasing flow. The motions of air and water organize the skies, the earth, and the oceans. The hidden logic of gases in turmoil paints great spinning eyes on the outer planets. Out of the collisions of particles in motion, desert dunes arise and hills become striped with sorted grains. Give these grains the ability to respond to their neighbours—make them fish, or birds, or buffalos—and there seems no end to the patterns that may appear, each an extraordinary collaboration that no individual has ordained or planned.

CONTENTS

Preface and Acknowledgements

After my 1999 book *The Self-Made Tapestry: Pattern Formation in Nature* went out of print, I'd often be contacted by would-be readers asking where they could get hold of a copy. That was how I discovered that copies were changing hands in the used-book market for considerably more than the original cover price. While that was gratifying in its way, I would far rather see the material accessible to anyone who wanted it. So I approached Latha Menon at Oxford University Press to ask about a reprinting. But Latha had something more substantial in mind, and that is how this new trilogy came into being. Quite rightly, Latha perceived that the original *Tapestry* was neither conceived nor packaged to the best advantage of the material. I hope this format does it more justice.

The suggestion of partitioning the material between three volumes sounded challenging at first, but once I saw how it might be done, I realized that this offered a structure that could bring more thematic organization to the topic. Each volume is self-contained and does not depend on one having read the others, although there is inevitably some cross-referencing. Anyone who has seen *The Self-Made Tapestry* will find some familiar things here, but also plenty that is new. In adding that material, I have benefited from the great generosity of many scientists who have given images, reprints and suggestions. I am particularly grateful to Sean Carroll, Iain Couzin, and Andrea Rinaldo for critical readings of some of the new text. Latha set me more work than I'd perhaps anticipated, but I remain deeply indebted to her for her vision of what these books might become, and her encouragement in making that happen.

Philip Ball

London, October 2007

THE MAN WHO LOVED FLUIDS

Leonardo's Legacy

Perhaps it is not so strange after all that the man who has come to personify polyvalent virtuosity, defining the concept of the Renaissance man and becoming a symbol for the unity of all learning and creative endeavour, was something of an under-achiever. That might seem an odd label to attach to Leonardo da Vinci, but the fact is that he started very little and finished even less. His life was a succession of plans made and never realized, of commissions refused (or accepted and never honoured), of studies undertaken with such a mixture of obsessive diligence and lack of system or objective that they could offer little instruction to future generations. This was not because Leonardo was a laggard; on the contrary, his ambitions often exceeded his capacity to fulfil them.

Yet if Leonardo did not achieve as much as we feel he might have done, that did not prevent his contemporaries from recognizing his extraordinary genius. The Italian artist and writer Giorgio Vasari was prone to eulogize all his subjects in his sixteenth-century *Lives of the Artists*, but he seems to make a special effort for Leonardo:

> In the normal course of events many men and women are born with various remarkable qualities and talents; but occasionally, in a way that transcends nature, a single person is marvellously endowed by heaven with beauty, grace, and talent in such abundance that he

leaves other men far behind, all his actions seem inspired, and indeed everything he does clearly comes from God rather than from human art. Everyone acknowledged that this was true of Leonardo da Vinci, an artist of outstanding physical beauty who displayed infinite grace in everything he did and who cultivated his genius so brilliantly that all problems he studied he solved with ease.

What Vasari did not wish to admit is that such an embarrassment of riches can be a burden rather than a blessing, and that it sometimes takes duller men to see a project through to its end while geniuses can only initiate them without cease. Leonardo's devotion to the study of nature and science could leave his artistic patrons frustrated. Isabella d'Este, marchesa of Mantua, was told by an emissary whom she dispatched to Florence to commission a portrait from the great painter, that 'he is working hard at geometry and is very impatient of painting . . . In short his mathematical experiments have so estranged him from painting that he cannot bear to take up a brush.'

But Leonardo was apt when the mood was upon him to labour without stint. His contemporary Matteo Bandello, a Piedmontese novelist, saw him at work on his ill-fated *Last Supper*: 'It was his habit often, and I have frequently seen him, to go early in the morning and mount upon the scaffolding . . . it was his habit, I say, from sunrise until dusk never to lay down his brush, but, forgetful alike of eating and drinking, to paint without intermission.' And yet his genius demanded space for reflection that he could ill afford. 'At other times', Bandello avers, 'two, three or four days would pass without his touching the fresco, but he would remain before it for an hour or two at a time merely looking at it, considering, examining the figures.' 'Oh dear, this man will never do anything!', Pope Leo X is said to have complained.

As his sketchbooks attest, lengthy and contemplative examination was his forte. When Leonardo looked at something, he saw more than other people. This was no idle gaze but an attempt to discern the very soul of things, the deep and elusive forms of nature. In his studies of anatomy, of animals and drapery, of plants and landscapes, and of ripples and torrents of water, he shows us things that transcend the naturalistic: shapes that we might not directly perceive ourselves but that we suspect we would if we had Leonardo's eyes.

We are accustomed to list Leonardo's talents as though trying to assign him to a university department: painter, sculptor, musician, anatomist, military and civil engineer, inventor, physicist. But his notebooks mock such distinctions. Rather, it seems that Leonardo was assailed by questions everywhere he looked, which he had hardly the opportunity or inclination to arrange into a systematic course of study. Is the sound of a blacksmith's labours made within the hammer or the anvil? Which will fire farthest, gunpowder doubled in quantity or in quality? What is the shape of corn tossed in a sieve? Are the tides caused by the Moon or the Sun, or by the 'breathing of the Earth'? From where do tears come, the heart or the brain? Why does a mirror exchange right and left? Leonardo scribbles these memos to himself in his cryptic left-handed script; sometimes he finds answers, but often the question is left hanging. On his 'to do' list are items that boggle the mind with their casual boldness: 'Make glasses in order to see the moon large.' It is no wonder that Leonardo had no students and founded no school, for his was an intensely personal enquiry into nature, one intended to satisfy no one's curiosity but his own.

We come no closer to understanding this quest, however, if we persist in seeing Leonardo as an artist on the one hand and a scientist and technologist on the other. The common response is to suggest that he recognized no divisions between the two, and he is regularly invoked to advertise the notion that both are complementary means of studying and engaging with nature. This doesn't quite hit the mark, however, because it tacitly accepts that 'art' and 'science' had the same connotations in Leonardo's day as they do now. What Leonardo considered *arte* was the business of making things. Paintings were made by *arte*, but so were the apothecaries' drugs and the weavers' cloth. Until the Renaissance there was nothing particularly admirable about art, or at least about artists— patrons admired fine pictures, but the people who made them were tradesmen paid to do a job, and manual workers at that. Leonardo himself strove to raise the status of painting so that it might rank among the 'intellectual' or liberal arts, such as geometry, music, and astronomy. Although a formidable sculptor himself, he argued his case by dismissing it as 'less intellectual': it is more enduring, admittedly, 'but excels in nothing else'. The academic and geometric character of treatises on painting at that time, most notably that of the polymath Leon Battista Alberti, which can make painting seem less a matter of inspiration than a process of drawing lines and plotting light rays, derives partly from this agenda.

Scienza, in contrast, was knowledge—but not necessarily that obtained by careful experiment and enquiry. Medieval scholastics had insisted that knowledge was what appeared in the books of Euclid, Aristotle, Ptolemy, and other ancient writers, and that the learned man was one who had memorized these texts. The celebrated humanism of the Renaissance did not challenge this idea but merely refreshed it, insisting on returning to the original sources rather than relying on Arabic and medieval glosses. In this regard, Leonardo was not a 'scientist', since he was not well schooled—the humble son of a minor notary and a peasant woman, he was defensive all his life about his poor Latin and ignorance of Greek. He believed in the importance of *scienza*, certainly, but for him this did not consist solely of book-learning. It was an *active* pursuit, and demanded experiments, though Leonardo did not exactly conduct them in the manner that a modern scientist would. For him, true insight came from peering beneath the surface of things. That is why his painstaking studies of nature, while appearing superficially Aristotelian in their attention to particulars, actually have much more of a Platonic spirit: they are an attempt to see what is really there, not what appears to be. This is why he had to sit and stare for hours: not to see things more sharply, but, as it were, to *stop* seeing, to transcend the limitations of his eyes.

Leonardo regarded the task of the painter to be not naturalistic mimicry, which shows only the surface contours and shallow glimmers of the world, but the use of reason to shape his vision and distil from it a kind of universal truth. 'At this point', Leonardo wrote of those who would grow tired of his theoretical musings on the artist's task, 'the opponent says that he does not want so much *scienza*, that practice is enough for him in order to draw the things in nature. The answer to this is that there is nothing that deceives us more easily than our confidence in our judgement, divorced from reasoning.' This could have been written by Plato himself, famously distrustful of the deceptions of painters.

I hope you can start to appreciate why I have placed Leonardo centre stage in introducing this volume of my survey of nature's patterns. As I explained in Book I, the desire to look *through* nature and find its underlying forms and structures is what characterizes the approach of some of the pioneers in the study of pattern formation, such as the German biologist Ernst Haeckel and the Scottish zoologist D'Arcy Wentworth Thompson. Haeckel was another gifted artist who firmly believed that the natural world needs to be arranged, ordered, tidied, before its forms and generative impulses can be properly perceived. Thompson shared Leonardo's conviction that the similarities of form and pattern we see in very different situations—for Leonardo it might be the cascades of a water spout and a woman's hair—reveal a deep-seated relationship. D'Arcy Thompson's view of such correspondences is one we can still accept in science today, based as it is on the idea that the same forces are likely to be at play in both cases. Leonardo's rationalization is more remote now from our experience, being rooted in the tradition of Neoplatonism that saw these correspondences as a central feature of nature's divine architecture: *as above, so below*, as the reductive formulation has it. When Leonardo calls rivers the blood of the Earth, and comments on how their channels resemble the veins of the human body, he is not engaging in some vague metaphor or visual pun; the two are related because the Earth is indeed a kind of living body and can therefore be expected to echo the structures of our own anatomy.

In this vision of a kind of hidden essence of nature, we can find the true nexus of Leonardo's 'art' and 'science'. We tend to think of his art as 'lifelike', and Vasari made the same mistake. He praises the vase of flowers that appears in one of Leonardo's Madonnas for its 'wonderful realism', but then goes on, I think inadvertently, to make a telling remark by saying that the flowers 'had on them dewdrops that looked more convincing than the real thing'. Leonardo might have answered that this was because he had indeed painted 'the real thing' and not what his eyes had shown him. His work is not photographic but stylized, synthetic, even abstract, and he admits openly that painting is a work not of imitation but of invention: 'a subtle *inventione* which with philosophy and subtle speculation considers the natures of all forms'. Leonardo 'is thinking of art not simply in technical terms', says art historian Adrian Parr, 'where the artist skillfully renders a form on the canvas . . . Rather, he takes the relationship of nature to art onto a deeper level, intending to express in his art "every kind of form produced in nature".' For indeed, as the art historian Martin Kemp explains, 'Leonardo saw nature as weaving an infinite variety of elusive patterns on the basic warp and woof of mathematical perfection.' And so, without a doubt, did D'Arcy Thompson.

LEONARDIAN FLOWS

While most painters used technique to create a simulacrum of nature, Leonardo felt that one could not imbue the picture with life until one understood how nature does it. His sketches, then, are not exactly studies but something between an experiment and a diagram—attempts to intuit the forces at play (Fig. 1.1). 'Leonardo's use of swirling, curving, revolving and wavy patterns, becomes a means of both investigating and entering into the rhythmic movements of nature', says Parr. Other western artists have tried to capture the forms of movement and flow, whether in the boiling vapours painted by J. M. W. Turner, the stop-frame dynamism of Marcel Duchamp's *Nude Descending a Staircase* (1912) or the fragmented frenzy of the Italian Futurists. But these are impressionistic, ad hoc and subjective efforts that lack Leonardo's scientific sense of pattern and order. Perhaps it is impossible truly to depict the world in this way unless you are a Neoplatonist. When John Constable declared in the early nineteenth

FIG. 1.1   A sketch of flowing water by Leonardo da Vinci.

century that 'Painting is a science and should be pursued as an inquiry into the laws of nature', he had in mind something far more mechanistic: that the painter should understand how physics and meteorology create a play of light and shadow, so that the paintings become convincing in an illusionistic sense.

But while a Leonardian perspective is valuable for surveying all nature's patterns, I have made him the pinion for this volume on patterns of motion, in fluids particularly, because there were few topics that enthralled him (and I mean that in its original sense) more than this. Of all the passions that he evinced, none seems more ardent than the wish to understand water. One senses that he regards it as the central elemental force: 'water is the driver of Nature', he says, 'It is never at rest until it unites with the sea . . . It is the expansion and humour of all vital bodies. Without it nothing retains its form.' It is no wonder, then, that one of Leonardo's most revealing and famous notebooks, known as the Codex Leicester or Codex Hammer,* is mostly concerned with water. There was hardly an aspect of water that Leonardo did not leave unexamined. He wrote about sedimentation and erosion in rivers, and how they produce meanders and sand ripples on the river bed (two patterns I consider later). He discussed how water circulates on the Earth in what we now call the hydrological cycle, evaporating from the seas and falling as rain on to high ground. He asked why the sea is salty and wondered why a man can remain underwater only 'for such a time as he can hold his breath'. He investigated Archimedes spirals for lifting water, as well as suction pumps and water wheels. He drew astonishing 'aerial' pictures of river networks (we'll see them in Book III), and planned great works of hydraulic engineering. In collaboration with Niccolò Machiavelli, he drew up a scheme to redirect the flow of the Arno River away from Pisa, thereby depriving the city of its water supply and delivering it into the hands of the Florentines.

It seems that Leonardo did not become fascinated by water because of his engineering activities; rather, according to art historian Arthur Popham, the latter were a symptom of the former: 'Something in the movement of water, its swirls and eddies, corresponded to some deep-seated twist in his

---

*The manuscript was acquired and published by Lord Leicester in Rome in the eighteenth century, but was bought in 1980 by the American Maecenas Armand Hammer.

nature.' No aspect of water captured his interest more than the eddies of a flowing stream. He wrote long lists of the features of these vortices that he intended at some point to investigate:

> Of eddies wide at the mouth and narrow at the base.
> Of eddies very wide at the base and narrow above.
> Of eddies of the shape of a column.
> Of eddies formed between two masses of water that rub together.

And so on—pages and pages of optimistic plans, of experiments half-started, of speculations and ideas, all described in such obsessive detail that even Leonardo scholars have pronounced them virtually unreadable. 'He wants', says the art historian Ernst Gombrich, 'to classify vortices as a zoologist classifies the species of animals.'

To judge from his sketches, Leonardo conducted a thorough, if haphazard, experimental programme on the flow patterns of water, watching it pass down channels of different shapes, charting the chaos of plunging waterfalls, and placing obstacles in the flow to see how they generated new forms. His drawings of water surging around the sides of a plate face-on to the flow show a delicately braided wake (Fig. 1.2a), and the resemblance to the braided hair of a woman in a preparatory study (Fig. 1.2b) is no coincidence, for as Leonardo said himself,

> Observe the motion of the surface of the water which resembles that
> of hair, which has two motions, of which one depends on the weight
> of the hair, the other on the direction of the curls; thus the water
> forms eddying whirlpools, one point of which is due to the impetus
> of the original current and the other to the incidental motion and
> return flow.

His self-portrait from 1512 shows his long hair and beard awash with eddies.

Many of these visual records are remarkably fine: he illustrates shock waves and ripples caused by constriction and widening of a channel (Fig. 1.3a), and his drawings of the flow past a cylindrical obstruction display the teardrop wake and the paired vortices (Fig. 1.3b) that have been found in modern experiments (see page 27). Fluid scientists today typically use techniques for revealing flow-forms that Leonardo is often said to have invented: fine particles that reflect light are suspended in the

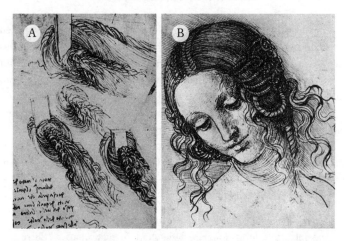

FIG. I.2 In the braided patterns of water flowing around a flat plate (a), Leonardo found echoes of the braids in a woman's hair (b).

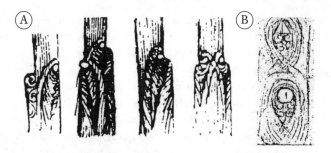

FIG. I.3 Leonardo sketched shock waves caused by constrictions in a channel (a) and the shape of wakes in flow around an obstacle (b).

water, or coloured dyes are added to part of the flow. 'If you throw sawdust down into a running stream', Leonardo said,

> you will be able to observe where the water turned upside down after striking against the banks throws this sawdust back towards the centre of the stream, and also the revolutions of the water and where other water either joins it or separates from it; and many other things.

Roughly speaking, these floating particles map out what are now called *streamlines*, which can be thought of as the trajectories of the flow.* In this sense, Leonardo's studies of flow patterns were thoroughly modern. But he had only his eyes and his memory to guide him in translating from what he saw to what he drew; and as art historians know, that translation occurs in a context of preconceived notions of style and motif that condition what is depicted. When Leonardo compares a flow to hair, he is struck initially by the resemblance, but then this correspondence superimposes what he knows of the way hair falls on what he sees in the stream of water. The result is, as Popham says, that although

> [t]he cinematographic vision which could see, the prodigious memory which could retain and the hand which could record these evanescent and intangible formations are little short of miraculous... [t]hese drawings do not so much convey the impression of water as of some exquisite submarine vegetable growth.

Was Leonardo able to do anything beyond recording what he perceived? Did he elucidate the reasons why these marvellous patterns are formed in water? If we have to admit that he did not really do that, it is no disgrace, since that problem is one of the hardest of all in physical science, and has still not been completely solved. On the whole the flows that Leonardo was studying were turbulent, fast-moving and unsteady in the extreme, so that they changed from one moment to the next. If he could describe these flows only in pictures and words, scientists could do no better than that until the twentieth century. And what vivid descriptions he gave!—

> The whole mass of water, in its breadth, depth and height, is full of innumerable varieties of movements, as is shown on the surface of

---

*A streamline has a technical definition: it is a line within the fluid for which the tangent at any point shows the direction of flow at that point. Streamlines show not only 'where the fluid is going' but also how fast: where streamlines are close together, the velocity is high. In steady flows, where the pattern of flow doesn't change over time, the path of a suspended particle or the trajectory of dye injected at a point, the particle path or so-called *streakline* of the dye trace out streamlines. But if the flow is unsteady, this is no longer true; the particle paths or streaklines can give an impression of the streamlines, and the true streamlines can be deduced from them, but they are not the same thing.

currents with a moderate degree of turbulence, in which one sees continually gurglings and eddies with various swirls formed by the more turbid water from the bottom as it rises to the surface.

Leonardo made some discoveries that stand up today. His comment that 'water in straight rivers is swifter the farther it is from the shore, its impediment', for example, is an elegant description of what fluid scientists call the velocity profile of flow in a channel, which is determined by the way friction between the fluid and the channel wall brings the flow there virtually to a standstill. Leonardo's explanation of how river meanders are caused by shifting patterns of sedimentation and erosion by the flow contain all the elements that today's earth scientists recognize.

His legacy for our understanding of fluid flow patterns goes deeper than this, however. As far as we can tell, Leonardo was the first Western scientist to really make the case that this phenomenon deserves serious study. And he showed that flowing water is not simply an unstructured chaos but contains persistent forms that can be recognized, recorded, analysed—forms, moreover, that are things of great beauty, of value to the artist as well as the scientist.

TRANSCENDENTAL FORMS

All the same, Leonardo's idiosyncratic, hermetic way of working meant that no research programme stemmed from his achievements. No scientist seems subsequently to have thought very much about fluid flows until the Swiss mathematician Daniel Bernoulli began to investigate them in the seventeenth century.*

Nor did Leonardo's work on fluid motion have any artistic legacy: his studies of flows as a play of patterns, forms, and streamlines leave no trace in Western art. Artists looked instead for a stylized realism which insisted that turbulent water be depicted as a play of glinting highlights and

*René Descartes made much of vortices, becoming convinced that the entire universe is filled with an ethereal fluid that swirls at all scales. Their gyrating motions, he said, carry along the heavenly bodies, explaining the circulations of the planets and stars. His theory, however, does not seem to owe any inspiration to Leonardo's work on eddies.

FIG. 1.4 *The Wreckers* by George Morland shows the typical manner in which Western painters depicted flow as a play of light. (Image: Copyright Southampton City Art Gallery, Hampshire, UK/The Bridgeman Art Library.)

surging foam: a style that is all surface, you might say. Just about any dramatic seascape of the eighteenth or nineteenth centuries will show this—George Morland's *The Wreckers* (1791) is a good example (Fig. 1.4).

A fluid style akin to Leonardo's does not show up again in Western art until the lively arabesques of the Art Nouveau movement of the late nineteenth century (Fig. 1.5). These artists took their inspiration from natural forms, such as the elegant curves and spirals of plant stems. As I discussed in Book I, the delicate frond-like forms discovered at this time in marine organisms and drawn with great panache and skill by Ernst Haeckel became a significant influence on the German branch of this movement, known as the Jugendstil—a two-way interaction that probably conditioned the way Haeckel drew in the first place. In England these trends produced something truly Leonardian in the works of the illustrator Arthur Rackham, where the correspondences between the waves and vortices of water, smoke, hair, and vegetation are particularly explicit

FIG. 1.5 Alphonse Mucha's Art Nouveau style emphasizes the arabesque patterns of flow.

(Fig. 1.6). But the use of vortical imagery here is really nothing more than a style, valued for its decorative and allusive qualities: there is no real sense that the artists are, like Leonardo, simultaneously conducting an investigation into nature's forms rather than simply adapting them for aesthetic ends.

One of the sources of the bold lines and sinuous forms of Art Nouveau is, however, more pertinent. In the mid-nineteenth century trade opened up between Western Europe and the Far East, and Japanese woodblock prints came into vogue among artist and collectors. Here Western artists found a very different way of depicting the world—not as naturalistic *chiaroscuro* but as a collage of flat, clearly delineated elements that disdains the rules of scientific optics and makes no pretence of photographic *trompe l'oeil*. To the Western eye these pictures are stylized and schematic, but some artists could see that this was not mere affectation, less still a simplification. What was being conveyed was the essence of things, unobstructed by superficial incidentals.

It is as simplistic to generalize about Chinese and Japanese art as it is about the art of the West—these traditions, too, have their different periods and schools and philosophies. But it is fair to say that most Chinese artists have attempted to imbue their works with *Ch'i*, the vital energy of the universe,

FIG. 1.6 Arthur Rackham's illustrations are Leonardian in their conflation of the eddies and tendrils of fluid flow and the swirling of hair. (Image: Bridgeman Art Library.)

the Breath of the Tao. *Ch'i* is undefinable and cannot be understood intellectually; the seventeenth-century painter's manual *Chieh Tzu Yüan* (The Mustard Seed Garden) explains that 'Circulation of the *Ch'i* produces movement of life.' So while the Taoist conviction that there exists a fundamental simplicity beyond the superficial shapes and forms of the world sounds Platonic, in fact it differs fundamentally. Unlike Plato's notion of static, crystalline ideal forms, the Tao is alive with spontaneity. It is precisely this spontaneity that the Chinese classical artist would try to capture with movements of the brush: 'He who uses his mind and moves his brush without being conscious of painting touches the secret of the art of painting', said the writer Chang Yen-yüan in the ninth century. In Chinese art everything depends on the brushstrokes, the source and signifier of *Ch'i*.

No wonder, then, that among the stroke types classified by artistic tradition was one called *T'an wo ts'un*: brushstrokes like an eddy or whirlpool. No wonder either that the ancient painters of China would

FIG. 1.7 In Chinese art, the flow of water is commonly represented as a series of lines approximating the trajectories of floating particles, like the streamlines employed by fluid dynamicists. This is not a 'realistic' but a schematic depiction of flow. These images are taken from a painting instruction manual compiled in the late seventeenth century. (From M. M. Sze (ed.) (1977), *The Mustard Seed Garden of Painting*. Reprinted with permission of Princeton University Press.)

say 'Take five days to place water in a picture.' What could be more representative of the Tao than the currents of a river swirling around rocks? But because the Tao is dynamic, an illusionistic rendering of a frozen instant, like that in Western art, would be meaningless. Instead, Chinese painters attempted to portray the inner life of flow, or what the twelfth-century Chinese critic Tung Yü called 'the fundamental nature of water'. They schematized flow-forms as a series of lines (Fig. 1.7), again remarkably like the scientist's streamlines. Some of Leonardo's sketches are very similar; one could almost mistake some of his drawings for those of an East Asian artist (Fig. 1.8).

FIG. 1.8 Some of Leonardo's sketches, such as this drawing of the Deluge, look remarkably 'East Asian'.

EBB AND FLOW

It is not quite true to say that Leonardo's project to animate his drawings of flow by capturing its fundamental forms and patterns has no parallels in Western art. Something like streamlines seem to resurface in Bridget Riley's early monochrome op-art paintings (Fig. 1.9), where the observer's eye is persuaded that there is real movement, real flow, still proceeding on the canvas. It may be that the *Spiral Jetty* (1970) of American earthwork artist Robert Smithson, a coil of rock and stone projecting into the Great Salt Lake of Utah, is meant to invoke one of the Leonardian vortices in the water that surrounds it. The American sculptor Athena Tacha makes extensive use of a vocabulary of flow forms that includes spirals, waves and eddies—her source of inspiration is made particularly explicit in a 1977 work *Eddies/Interchanges (Homage to Leonardo)* (Fig. 1.10), which she proposed as a walkway or even a 'drive-in sculpture'.

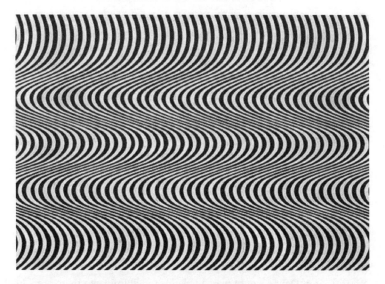

FIG. 1.9 Many of Bridget Riley's early op-art paintings, such as *Current* (1964), show something akin to streamlines that convey a genuine sense of movement.

FIG. 1.10 Athena Tacha's *Eddies/Interchanges (Homage to Leonardo)* (1977). The sculpture exists only as a maquette, but was intended to be made on a large scale. (Photo: Athena Tacha.)

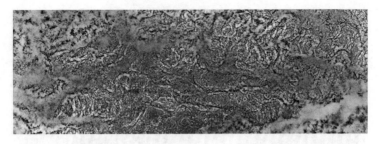

FIG. 1.11   The fleeting forms of turbulent fluid flow in the River Taw in south-west England were captured in night-time photographs by the artist Susan Derges. (Photo: Susan Derges.)

But perhaps the modern works that most successfully recapitulate Leonardo's enquiry into the forms of nature are those of the British photographer Susan Derges. She immersed huge sheets of photographic paper protected between glass plates just beneath the water surface of the River Taw in Devon, south-west England, and illuminated them at night with a very brief flash of light. All the little peaks and troughs of surface waves are imprinted on the photographic image as a kind of shadowgraph (Fig. 1.11). Overhanging vegetation is sometimes imprinted too, evident only as a silhouette in the manner of a Japanese print. Derges has herself studied Japanese art—she lived in Japan in the 1980s, where she was influenced by the works of Hiroshige and Hokusai—and she is familiar with the Taoist notion of distilling the universal from the particular.

Like Leonardo's drawings, these photographs could serve either as works of art or as scientific records, since what emerges from a dialogue with nature's patterns can be viewed either way.

PATTERNS DOWNSTREAM

Order That Flows

There is nothing new in the idea that the transient forms of fluid flow, frozen by the blink of a camera's shutter, have artistic appeal. As early as the 1870s, the British physicist Arthur Worthington used high-speed photography to capture the hidden beauty of splashes. He dropped pebbles into a trough of water and discovered that the splash has an unguessed complexity and beauty with a surprising degree of symmetry and order. Worthington worked at the Royal Naval College in Devonport on the south-west coast of England, where the study of impacts in water had decidedly unromantic implications; but one senses that Worthington lost sight of the military origins of his research as he fell under the allure of these images. A splash, he found, erupts into a corona with a rim that breaks up into a series of spikes, each of them releasing micro-droplets of their own (Fig. 2.1). There is, he said, something seemingly 'orderly and inevitable' in these forms, although he admitted that 'it taxes the highest mathematical powers' to describe and explain them. In 1908 he collected his pictures in a book called *A Study of Splashes*, which aimed to please the eye as much as to inform the mind.

Worthington realized that the images were clearest if the liquid was opaque, and so he used milk instead of water (the two are not equivalent, for the higher viscosity of milk alters the shape of the splash). His sequences of photos appear to represent a series of successive snapshots

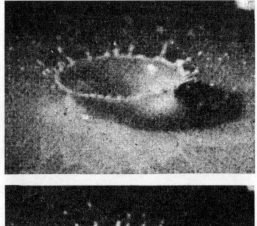

FIG. 2.1 The splash of a drop of milk, photographed by Arthur Worthington in the late nineteenth century.

taken during the course of a single splash. But that is a forgivable decep-
tion, for Worthington didn't have a camera shutter able to open and close
at such a rate. Instead, each splash yielded a single image, revealed in a
flash of light from a spark that lasted for just a few millionths of a second in

a darkened room. To capture the sequence, Worthington simply timed the spark at successively later instants in the course of many splashes that he hoped were more or less identical.

Worthington suspected that these pictures might appeal to the public's sense of beauty, but he didn't capitalize on them with anything like the chutzpah of his successor, the American electrical engineer Harold Edgerton of the Massachusetts Institute of Technology. In the 1920s Edgerton realized that the newly invented stroboscope could 'freeze' rapid, repetitive motions when the flash rate of the lamp was synchronized with the cycling rate of the movement. He developed a stroboscopic photographic system that could take 3,000 frames per second. His high-speed photographs became famous thanks to Edgerton's sense of eye-catching subject and composition: he took split-second pictures of famous sportspeople and actors, and his iconic 'Shooting the Apple' pays homage to the legend of William Tell while revealing the compelling destructive-ness of a speeding bullet. In Edgerton's quickfire lens, water running from the tap becomes petrified into what appears to be a mound of solid glass. His book *Flash* (1939) was unashamedly populist, a coffee-table collection of remarkable shots, and his film *Quicker 'n a Wink* (1940) won an Oscar the following year for the best short film.

But probably the most memorable of Edgerton's images was copied straight from Worthington: he filmed milk droplets as they splash into a smooth liquid surface. Edgerton's drop is tidier, somehow more regular and orderly, a true marvel of natural pattern (Fig. 2.2a): each prong of the crown is more or less equidistant from its neighbours, and each of them disgorges a single spherical globule.* This is the secret structure of rainfall, reproduced countless times as raindrops fall into ponds and puddles. Edgerton's milk splash has become an icon of hidden order, as much a work of art as a scientific study. More prosaically, the image was adopted in the 1990s in stylized form by the British milk-marketing and distribution company Milk Marque (Fig. 2.2b). D'Arcy Thompson was captivated by these structures,

---

*You can watch Edgerton's film of the splash online at <http://web.mit.edu/edgerton/ spotlight/Spotlight.html>. One can hardly view it now without noting the chilling resem-blance to the aerial footage of the hydrogen-bomb tests of the 1950s—a documentary technology to which Edgerton himself contributed.

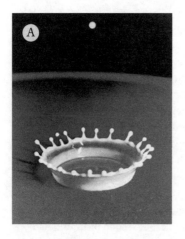

Milk Marque

FIG. 2.2 Harold 'Doc' Edgerton's milk splash, photographed at MIT, is tidier than Worthington's, and reveals more of the structure's symmetry (a). This iconic image was used in a stylized form by a British milk-marketing company in the 1990s (b). (Photo a: Edgerton Center, Massachusetts Institute of Technology.)

too. In his classic book *On Growth and Form* (1917) he compared Worthington's fluted cup with its 'scolloped' and 'sinuous' edges to the forms a potter makes at a more leisurely pace from wet clay. Edgerton's photograph provided the front plate for the 1944 revised edition of the book, where it was as though Thompson were saying 'Look here, *this* is my subject. Here is the full mystery—the quotidian, ubiquitous mystery—of pattern.'

To Thompson, who possessed a finely honed instinct for similarities of pattern and shape in nature, these splash-forms were not just a curiosity of fluid flow but a manifestation of a more general patterning process that could be seen also in the shapes of soft-tissued living organisms. The bowl-like structure with its notched rim, he said, is echoed in some species of hydroid, marine animals related to jellyfish and sea anemones (Fig. 2.3). Of course, here the form is persistent, not literally gone in a flash; yet 'there is nothing', Thompson said, 'to prevent a slow and lasting manifestation, in a viscous medium such as a protoplasmic organism, of phenomena which appear and disappear with evanescent rapidity in a more mobile liquid.' These organisms, he argued 'might conceivably display configurations

22

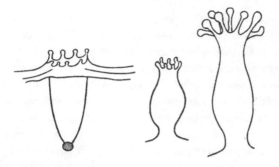

FIG. 2.3 D'Arcy Thompson noted similarities between Worthington's splashes, sketched on the left, and a type of hydroid, on the right.

analogous to, or identical with, those which Mr Worthington has shewn us how to exhibit by one particular experimental method.'

As is often the case with *On Growth and Form*, the argument here is largely a matter of wishful thinking. There is no good reason to think that a hydroid grows as a splash grows—why, after all, should we then expect it to become arrested in a particular 'snapshot', and not to erupt, fragment and subside like the droplets that Worthington and Edgerton produced?

All the same, there is a pattern here that demands explanation. What is the reason for the undulating corona of a splash? Surprisingly, this is still not clear. Whatever else, it is evidently a symmetry-breaking process, for the droplet initially has perfect circular symmetry when seen from above; but in the corona this is disrupted as the spikes appear. What is more, the process somehow introduces a characteristic distance or *wavelength*: the separation between adjacent spikes is more or less constant around the rim. Below we will see other examples of this 'wavelength selection' in the patterns of fluid flow.

WHORLS APART

The splash is an oddity, a curious little quirk of fluid behaviour. If one were to judge from Leonardo's studies, one might conclude that the leitmotif of fluid flow is a different structure, less symmetrical but still exhibiting an unmistakable sense of organization: the whirlpool or vortex (Fig. 2.4). When you think about it, the vortex is stranger and more unexpected than the splash corona. The latter is a classic example of a broken symmetry, a circle that develops a wobble; but the vortex seems to come out of nowhere. Think of a river, flowing gracefully down a barely perceptible gradient: why should the water start suddenly to deviate sideways, where no gradient seems to drive it, and then—curioser and curioser—to circulate *back* on itself, flowing (or so it seems) *uphill*? Whence this apparently irrepressible tendency of a liquid to swirl and coil?

What this question calls for is a science of fluid flow. That discipline is variously called hydrodynamics (an indication of how water-centred the topic is), fluid mechanics and fluid dynamics. I am going to explain something about its theoretical foundation in this book's final chapter; but let me admit right now that this is not going to be particularly revelatory. The theory of fluid dynamics is rather simple in conception, unutterably difficult in most applications (unless you have help from a

FIG. 2.4 Leonardo seemed to consider the vortex to be a fundamental feature of fluid flow.

24

powerful computer), and of limited value in providing any kind of intuitive picture of why fluids possess such an unnerving propensity for pattern. It is, furthermore, a theory that is incomplete, for we still lack any definitive understanding of the most extreme yet also the most common state of fluid flow, which is turbulence. In everyday parlance, 'turbulent' is often a synonym for the disorganized, the chaotic, the unpredictable—and while fluid turbulence does display these characteristics to a greater or lesser degree, we can see from Leonardo's sketches (which invariably show turbulent flows) that there is a kernel of orderliness in this chaos, most especially in the sense that turbulent flow often retains the organized motions that spawn vortices.

For now, I shall describe fluid flow in the manner in which scientists since Leonardo have been mostly compelled to do: by observing and drawing pictures and writing not equations but prose. The French mathematician Jean Leray, one of the great pioneers of fluid dynamics in the twentieth century, formulated his ideas while gazing for long hours at the problem in hand, standing on the Pont Neuf in Paris and watching the Seine surge and ripple under the bridge. It is a testament to Leray's genius that this experience did not simply overwhelm him, for, as much as you may plot graphs and make meticulous lab notes, observing the flow of fluids can easily leave you with a sense of grasping at the intangible.

Thinking about the problem as Leray did can at least help us to see where we should start. Here is the Seine—not, by all accounts, the most sanitary of rivers in the early part of the last century—streaming around the piles of the Pont Neuf. The water parts as it flows each side of the pillars, and this disturbance leaves it billowing and turbulent downstream. To use the terminology we encountered in the first chapter, the streamlines become highly convoluted. How does that happen? Let's back up a little. If the water were not moving at all—if, instead of a river, the pillar stands in a stagnant pond—then there is no pattern, since there is no motion and no streamlines. We must ask how still, uniform water becomes eddying flow. Let's turn on the flow gradually and see what happens.

So here, then, is our idealized Seine: water flowing down a shallow channel, which for simplicity we will assume to be flat-bottomed with parallel, vertical sides. At slow flow rates, all the streamlines are straight and parallel to the direction of flow—in other words, any little particle that traces the flow, such as a leaf floating on the river surface, will follow a simple, straight trajectory (Fig. 2.5a). At the edge of the 'river', where the fluid rubs against the confining walls, we can imagine that something more complicated might happen, but actually this need not alter the picture very much,* and in any event we can ignore that if we make the river wide and focus on the middle, where the streamlines are parallel and all of the fluid moves in synchrony, in the same direction at the same speed. Flows like this in which the streamlines are parallel are said to be *laminar*. The flow here is uniform throughout the water's depth (again, we can ignore the region where the water drags against the river bottom), so we can depict it simply in terms of two-dimensional streamlines.

Now it's time to introduce the Pont Neuf, or, rather, the scientist's idealized version, which is a single cylindrical column standing in the middle of the river (Fig. 2.5b). Clearly, some streamlines have to be deflected around the cylinder. If the flow rate is very low, this can happen smoothly: the streamlines part as they reach the cylinder and converge again downstream to restore the laminar flow (Fig. 2.5b,c). This creates a contained, lens-shaped region of disruption.

What happens if the flow rate increases? In the wake of the pillar, we find that two little counter-circulating vortices, or eddies, appear (Fig. 2.5d). The streamlines in the eddies are closed loops: there are little pockets of fluid that have become detached from the main flow and remain in place behind the pillar. A particle carried along in the water would go round and round for ever if it got trapped in these eddies. If the flow rate becomes still greater, the eddies grow and get stretched out (Fig. 2.5e), but the flow outside of them remains laminar: the streamlines eventually converge downstream and resume their parallel paths.

It would be handy to have some way of measuring when these changes happen. But we cannot simply say that, for example, the pair of eddies appear at a flow speed of ten centimetres per second or whatever, because

---

*A fluid will typically be slowed by friction where it touches the wall. If we think of it as being divided into a series of parallel strips, the outermost strip may be arrested entirely by friction. The strip next to it is slowed down by the motionless layer, but not stopped entirely; and so on with successive strips, each slowed a little less. So the velocity of the flow increases smoothly from the walls, where it is zero, to the middle of the flow—just as Leonardo said.

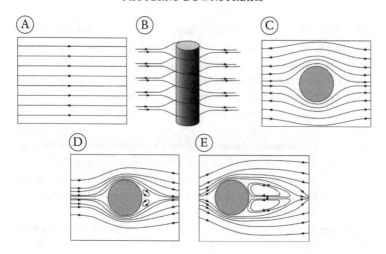

FIG. 2.5 Streamlines in the river. When fluid flow is slow and undisturbed, a floating particle follows a straight-line trajectory (a). But if an obstacle is placed in the stream, the water must pass around it to either side (b). At low flow speeds, the flow remains identical in all perpendicular layers, and so it can be represented in a single flat plane (c). Upstream, the diverted streamlines converge again. But at faster speeds, circulating vortices appear behind the obstacle (d), which grow and elongate as the flow quickens its pace (e).

in general this threshold also depends on factors other than the fluid velocity, in particular the width of the pillar and the viscosity of the liquid. However, one of the most profound and useful discoveries of fluid dynamics is that flows can be described in terms of 'universal' measures that take all of these things into account. In this case, let us assume that the flow is happening in a channel so wide that the banks are very distant from the pillar and have no effect on the flow there. Then we find that the flow velocity at which eddies first appear, multiplied by the diameter of the pillar and divided by the viscosity of the liquid, is always a constant, regardless of the type of liquid or the dimensions of the pillar. This number has no units—they all cancel out—but simply has a value of about four.

This is an example of a *dimensionless number*, one of the 'universal parameters' that allows us to generalize about fluid flows without having to take into account the specific details of our experimental system. It is called the

Reynolds number, after the British scientist Osborne Reynolds who studied fluid flow in the nineteenth century. It isn't just a happy coincidence that this particular combination of experimental parameters eliminates all units and gives a bare number. Dimensionless numbers in fluid dynamics are in fact *ratios* that express the relative contributions of the forces influencing the flow. The Reynolds number (*Re*) measures the ratio of the forces driving the flow (quantified by the flow velocity) to the forces retarding it by viscous drag. In our experiment, the size of the pillar and the liquid viscosity stay constant, and so *Re* increases in direct proportion to the flow speed.

So then, at a Reynolds number of four, the flow pattern changes abruptly with the appearance of the pair of vortices. The new pattern remains, albeit with increasingly elongated vortices, until *Re* reaches a value of about 40. Then something new happens: the downstream streamlines don't all eventually become parallel, but instead there is a persistent wavy undulation in the wake. This can be seen experimentally by injecting a coloured dye into the liquid from the rear side of the cylinder, which is carried along in a narrow jet that more or less mirrors the streamlines (Fig. 2.6*a*). As the Reynolds number (that is, the flow rate) continues to increase, the waves get more pronounced, and the peaks become sharper (Fig. 2.6*b*). Around *Re* = 50, these crests break and curl over into swirling vortices (Fig. 2.6*c*): an astonishing and very beautiful pattern in which we can immediately see the characteristic traceries of Art Nouveau. In effect, the wake of the flow is continually shedding eddies, first on one side and then on the other.

Although, as we have seen, structures rather like this can be seen in Leonardo's sketches, they do not seem to have been reported in a formal scientific context until 1908, when the French physicist Henri Bénard published a paper called 'Formation of rotation centres behind a moving obstacle'. But Bénard's work was not known to the German engineer Ludwig Prantl when he made a study of cylinder wakes in 1911. Prantl had a theory for such flows, and the theory said that the wake should be smooth, rather like that in Fig. 2.5*c*. But when his doctoral student Karl Hiemenz conducted experiments on this arrangement, he found that the flow behind the obstacle underwent oscillations. Nonsense, Prantl told him—clearly the cylinder isn't smooth enough. Hiemenz had it repolished, but found the same result. 'Then your channel is not perfectly symmetrical,' Prantl told his hapless student, forcing him to make further improvements.

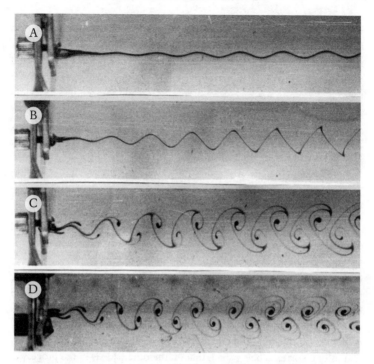

FIG. 2.6 At a flow rate corresponding to a Reynolds number of about 40, the wake of a flow past a cylinder develops a wavy instability, revealed here by the injection of a dye. At higher flow speeds this wavy disturbance develops into a train of vortices, called a Kármán vortex street (*c,d*). Above a Reynolds number of around 200, the vortex street breaks up into a turbulent wake. (From Tritton, 1988).

At that time, a Hungarian engineer named Theodore von Kármán came to work in Prantl's laboratory in Göttingen. He began to tease Hiemenz, asking him each morning, 'Herr Hiemenz, is the flow steady now?' Hiemenz would sigh glumly, 'It always oscillates'. Eventually, von Kármán decided to see if he could understand what was going on. A talented mathematician, he devised equations to describe the situation and found they predicted that vortices behind the cylinder could be stable. As a result of this work, the trains of alternating vortices—which, contrary to Prantl's suspicions, are a real and fundamental feature of the flow—are now known as Kármán vortex streets.

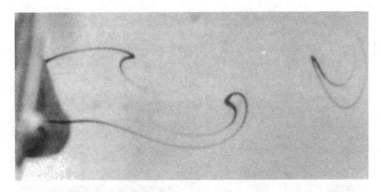

FIG. 2.7 The Kármán vortex street arises from 'eddy shedding'. Circulating vortices behind the obstacle are shed from alternate sides and borne along in the wake. Here one eddy is in the process of forming just after that on the opposite side has been shed. (From Tritton, 1988.)

Where do the vortices come from? They spring out of the layer of fluid moving past the surface of the cylinder, which acquires a rotating tendency called vorticity from the drag induced by the obstacle. This process is highly coordinated between the 'left' and 'right' sides of the pillar, so that as one vortex is being shed, that on the other side is in the process of forming (Fig. 2.7). Vortex streets are common in nature. They have been seen imprinted on clouds as air streams past some obstacle such as a region of high pressure (Fig. 2.8). They are generated in the wake of a bubble rising through water, pushing the bubble first to one side and then the other as the vortices are shed; this explains why the bubbles in champagne often follow a zigzag path as they rise. Vortex shedding from the wingtips of flying insects helps them to defeat the usual limitations of aerodynamics: in effect, the insects rotate their wings after a downstroke so that they receive a little push from the circulating eddy this creates.

If the flow rate is increased still further, the vortices in the street begin to lose their regularity, and the wake of the pillar seems to degenerate into chaos. But in fact the orderliness of the flow comes and goes: an observer stationed downstream would see more or less orderly vortex streets pass by, interrupted now and then by bursts of disorderly

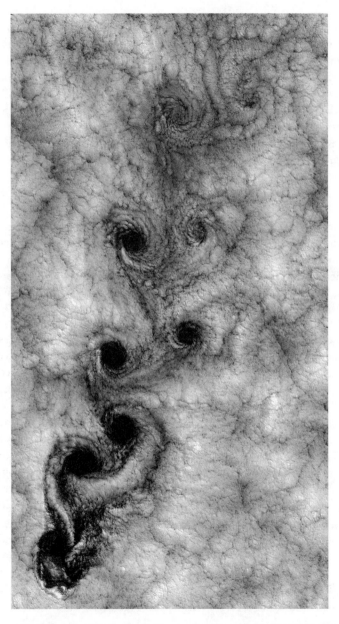

FIG. **2.8** A vortex street in clouds due to disruption of an atmospheric flow. (Photo: NOAA/University of Maryland Baltimore County, Atmospheric Lidar Group.)

turbulence. Above $Re = 200$, however, an observer a long distance downstream would note that the ordered vortex patterns seem to have vanished for good. Even then, vortex streets persist close to the pillar itself, but they get scrambled as they move downstream. At $Re = 400$, however, even this organization gets lost and the wake looks fully turbulent. This is the typical situation for a river passing around the piles of a bridge—rivers generally have a Reynolds number of more than a million—and so Leray will have strained in vain to discern much of a pattern in the murky Seine.

UNSTABLE ENCOUNTERS

The transformation of a smooth, laminar flow into the wavy pattern shown in Fig. 2.6a illustrates a common feature of pattern-forming systems: the sudden onset of a wobble when the system is driven hard enough. I discussed several such wave-like instabilities in Book I, from the fragmentation of columns of liquid to the appearance of oscillations in a chemical reaction. What creates the wave in this instance?

It is an example of a *shear instability*. When two layers of fluid slide past one another, they rub against each other and experience a so-called shear force. In the tail of the wake immediately behind the pillar, the fluid flow is slowed down, for here the flow is impeded. In the same way, swimmers forced to manoeuvre around a lane-blocking obstacle will take longer to reach the far end of the pool than identical competitors whose lanes are clear. This means that adjacent layers of fluid move at different speeds, and so there is a shear force at the boundary. This can amplify ripples that develop here by chance.

The situation is clearer if we consider adjacent layers of liquid flowing not just at different speeds, but in opposite directions.* Imagine that a bulge appears at the interface. Where the bulge pushes out into the next layer, the liquid there is 'squeezed' and flows faster, just as a river flows more rapidly if it enters a narrow gorge. Meanwhile, the bulge widens the

---

*That may sound like a very different case, but in fact it isn't. In the former case, it appears from the perspective of the faster layer that the slower layer is going backwards, just as a car that you've overtaken seems to be receding into the distance behind you.

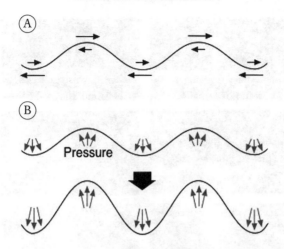

FIG. 2.9   In a shear flow where two layers of fluid move past each other, the boundary is susceptible to a wavy instability. On the concave side of a bulge, the flow is slowed down, while on the convex side it is speeded up (*a*). This causes a difference in pressure that pushes the bump outwards, amplifying it (*b*). Eventually these waves peak and curl into vortices. This is called the Kelvin–Helmholtz instability.

layer from which it emerges, and the flow there is slowed like a river becoming more broad and sluggish when it empties onto a wide flood plain (Fig. 2.9). In 1738 Daniel Bernoulli showed that the pressure exerted by a liquid lateral to the direction of flow decreases as the flow gets faster. This explains why the shower curtain always sticks to you: as the jet of water moves the layer of air between your skin and the curtain, the pressure there falls and the curtain gets pushed inwards by the air pressure on the other side.

This means that there is low pressure on the convex side of the bulge and high pressure on the concave side, so the bulge gets pushed outwards and accentuated. In other words, there is positive feedback: the more the bulge grows, the greater its tendency to grow further. This seems to imply that *any* bulge at the boundary of a shear flow will be self-amplifying. But in practice, the viscosity of the liquid (a measure of its resistance to flow) damps out the instability until the shear force (here depending on the relative velocity of the two layers) exceeds some critical threshold. What is more, the

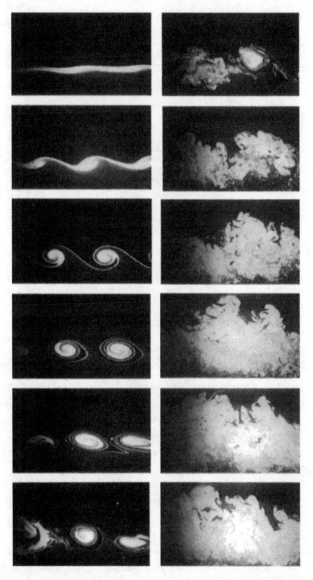

FIG. 2.10 Evolution of a Kelvin–Helmholtz instability in a shear flow, made visible by injecting a fluorescent dye at the boundary of the two flows. The waves roll over into vortices, which then interact and break up into turbulence. The sequence progresses from top to bottom, first on the left and then on the right. (Photo: Katepalli Sreenivasan, Yale University.)

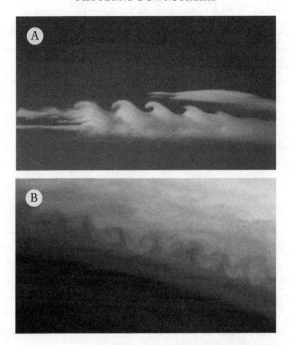

FIG. 2.11 A Kelvin–Helmholtz instability in atmospheric clouds (*a*), and in the atmosphere of Saturn (*b*). (Photos: *a*, Brooks Martner, NOAA/Forecast Systems Laboratory; *b*, NASA.)

self-amplification is greatest at a particular wavelength of undulation, and so this wavy pattern gets 'selected' from all the others. The result is that the shear flow develops a regular series of waves (Fig. 2.10).

This shear instability was studied in the nineteenth century by two of its greatest physicists, Lord Kelvin and Hermann von Helmholtz, and it is now known as the Kelvin–Helmholtz instability. The waves become sharply peaked as the structure evolves, and are then pulled over into curling breakers, producing a series of vortices.* Kelvin–Helmholtz instabilities are another of the patterning mechanisms that operate in the

---

*I must stress that this is *not* how the vortices of a Kármán vortex street are made, however. The waviness in Fig. 2.6 is indeed a shear instability, but the vortices grow from the edge of the pillar, not from the peaks of the downstream waves.

atmosphere, appearing for example in clouds or air layers (Fig. 2.11a). I have seen them myself in the sky above London. NASA's Cassini spacecraft captured a particularly striking example in the atmosphere of Saturn, where bands of gases move past one another (Fig. 2.11b).

PLUGHOLES AND WHIRLPOOLS

Shear instabilities can thus stir fluids into whirlpools. These flow-forms range in scale from the mundane spiralling of bath water around the plughole to the terrifying gyrations of tornadoes and hurricanes (Fig. 2.12). The bathtub vortex puzzled scientists for centuries. Leonardo described it: the vortex, he said, 'will be bored through down to the hole's outlet; and this hollow will be filled with air down to the bottom of the water'. He asserted that this vortex must be a transient phenomenon, because water is heavier than air and so the walls must eventually collapse.

Where does the rotation come from? In 1955 a French hydraulic engineer named Francis Biesel wrote that the slightest rotational circulation 'diffused throughout the fluid mass' could become concentrated in the funnel-shaped outflow. 'Experiment indicates that it is an eminently unpredictable phenomenon', he wrote. 'It is also a particularly persistent one, quite difficult to counter.' But if there's no rotation there to begin with, he said, it cannot be created from nothing.

A popular notion says that the rotation of the earth starts the bathtub vortex spinning. But while it is certainly true that this rotation controls the direction of the giant atmospheric vortices of cyclones, which rotate counter-clockwise in the Northern Hemisphere and clockwise in the Southern, the influence of the Earth's rotation on a micro-cyclone in the bath should be extremely weak. Biesel claimed that it cannot be responsible for the bathtub vortex because, contrary to popular belief, they may rotate in either direction at any place on the planet.

But is that really so? In 1962 the American engineer Ascher Shapiro at the Massachusetts Institute of Technology claimed that he had consistently produced counter-clockwise vortices in his lab by first allowing the water to settle for 24 hours, dissipating any residual rotational motion, before pulling the plug. The claim sparked controversy: later researchers

FIG. 2.12 Vortices in fluids occur on many scales, from bath plugholes to marine whirlpools (*a*) to hurricanes (*b*). (Photos: *b*, NASA.)

said that the experiment was extremely sensitive to the precise conditions in which it was conducted. The dispute has never quite been resolved.

We do know, however, why a small initial rotation of the liquid develops into a robust vortex. This is due to the movement of the water as it converges on the outlet. In theory this convergence can be completely symmetrical: water moves inwards to the plughole from all directions. But the slightest departure from that symmetrical situation, which could happen at random, may be amplified because of the way fluid flow operates. Flow may be transmitted from one region of fluid to another because of friction. This is why you can stir your coffee by blowing across the top, and why ocean surface currents are awakened by the wind: one flow drives another. A small amount of rotation excites more, and then more again ... To sustain this process, however, the nascent vortex needs to be constantly supplied with momentum, just as you need to keep pushing a child on a swing to keep them moving. This momentum is provided by the inflow of water towards the plughole: in effect, the momentum of movement in a straight line is converted to the momentum of rotation.

The plughole vortex is an example of spontaneous symmetry-breaking: a radially converging flow, with circular symmetry, develops into a flow with an asymmetric twist, either clockwise or counter-clockwise depending on the nature of the imperceptible push that gets the rotation under way. Setting aside Shapiro's ambiguous experiments, this initial kick seems to happen at random, and there is no telling which way the bathtub vortex will spin.

Marine whirlpools have spawned many legends, from Charybdis of the *Odyessy* to the Maelstrom of Nordic tales. Centrifugal forces act on the spinning water to push the surface of a whirlpool into an inverted bell-shape, which is embellished by ripples excited near the centre to produce the familiar corkscrew appearance (Fig. 2.12*a*). Some of these structures, like the Maelstrom and the vortex at St Malo in the English Channel, are caused by tidal flows near to shore, which is precisely why they are so hazardous to seamen. Poe's terrifying account of a Norwegian fisherman's *Descent into the Maelström* ('the boat appeared to be hanging, as if by magic, midway down, upon the interior surface of a funnel vast in circumference, prodigious in depth') is uncannily accurate not just in a pictorial sense but in terms of the underlying fluid dynamics, suggesting that perhaps Poe took his information from a real-life encounter.

Vortices appear not just in sluggishly flowing fluids but also in fully turbulent ones. Although such flows seem disorderly and unpredictable, nevertheless the fluids retain a propensity to organize themselves into these distinct, coherent structures. This was demonstrated by Dutch physicists GertJan van Heijst and Jan-Bert Flór at the University of Utrecht, who showed that a kind of two-headed vortex (the technical term is 'dipolar') can emerge from a turbulent jet. They fired a jet of coloured dye into water whose saltiness increased with depth. This gradient in saltiness meant that the water got denser as it got deeper, which suppressed up-and-down currents in the fluid, making the flow essentially two-dimensional: each horizontal layer flowed in the same way. The initially disordered flow in the head of the jet gradually arranged itself into two counter-rotating lobes (Fig. 2.13). And to show just how robust these dipolar vortices are, van Heijst and Flór fired two of them at each other from opposite directions, so that they collided head on. This might be expected to generate a turbulent frenzy, but instead the vortices displayed a slippery resilience that somehow makes you think of egg yolks. As they collided, they simply paired up with their counterpart in the other jet and, without mixing, set off in new directions (Plate 1).

THE GIANT'S EYE

One of the most celebrated and dramatic of the vortices found in a natural turbulent flow has been gyrating for over a century. Jupiter's

FIG. 2.13   A turbulent jet injected into a stratified fluid (in which a density gradient keeps the flow essentially two-dimensional) organizes itself into a coherent structure, the dipolar vortex. (Photos: GertJan van Heijst and Jan-Bert Flór, University of Utrecht.)

Great Red Spot is a maelstrom to cap them all: as wide as the Earth and three times as long, it is a storm in Jupiter's southern hemisphere in which the winds reach speeds of around 350 miles an hour (Plate 2). It is often said that the Red Spot was first observed in the seventeenth century by Robert Hooke in England and by Giovanni Domenico Cassini in Italy. But it is not clear that either of these scientists saw today's Red Spot. Cassini's spot, which he reported in 1665, was subsequently observed until 1713, but after that the records fall silent until the sighting of the present Red Spot in 1830. Vortices like this do come and go on Jupiter—three white spots to the south of the Great Red Spot appeared in 1938 and persisted until 1998, when they merged into a single spot.*
The Great Red Spot itself seems to be diminishing since its observation in the nineteenth century, and it is very likely that one day Jupiter's eye will

*In 2005–6 this turned red, presumably because its increased strength dredged up some red material from deeper in the atmosphere. It has been dubbed Red Spot Junior.

close again. How do these structures arise, and how can they for so long defy the disruptive pull of turbulence?

The colours of Jupiter's cloudy upper atmosphere are caused by its complex chemical make-up: a mixture of hydrogen and helium with clouds of water, ammonia, and other compounds. All this is stirred by the planet's rotation into a swirling brew, which is patterned even before we start to consider the spots. The Jovian atmosphere is divided into a series of bands marked out in different colours (Plate 3). Each band is a 'zonal jet', a stream that flows around lines of latitude either in the same or the opposite direction to the planet's rotation. The Earth has zonal jets, too: the westward current of the trade winds in the tropics, and the eastward current of the jet stream at higher latitudes. On Jupiter, both hemispheres have several zonal jets travelling to the east and west. The origin of these bands is still disputed, but they may be the product of small-scale eddies pulled and blended into latitudinal jets by the planet's rotation.

Peter Olson and Jean-Baptiste Manneville have shown that a similar banded structure can arise from convection in a laboratory model of Jupiter's atmosphere. They used water to mimic the fluid atmosphere (since it has a similar density), trapped between two concentric spheres 25 and 30 centimetres across. The inner sphere was chilled by filling it with cold antifreeze; the outer sphere was of clear plastic, so that the flow pattern could be seen. The researchers simulated the effect of the planet's gravity by spinning both spheres to create a centrifugal force, and added a fluorescent dye to the water so that the flow pattern could be seen under ultraviolet light. They saw zonal bands appear around their model planet because of convective motions. We will see in the next chapter how such rolls and stripes are a common feature of convection patterns.

The spot features in Jupiter's atmosphere are formed at the boundary of two zonal jets, where the movement of gases in opposite directions creates an intense shear flow. The Great Red Spot circulates like a ball bearing between the flows above and below (Fig. 2.14). There is now reason to think that a single big vortex like this may be a very general feature of this kind of turbulent flow. Philip Marcus from the University of California at Berkeley has carried out numerical calculations of the flow in a thin annulus of fluid: a washer-shaped disk

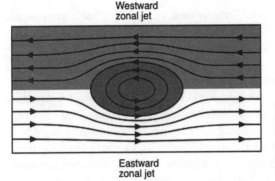

**Westward
zonal jet**

**Eastward
zonal jet**

FIG. 2.14 Jupiter's
Great Red Spot
circulates between
oppositely directed
zonal jets that
encircle the planet.

with a hole in the middle, representing a kind of two-dimensional projection of one of Jupiter's hemispheres. The rotation itself sets up a shear flow: rings of fluid at successively larger radial distances from the centre flow past each other. Marcus found that when the shearing was high enough to make the flow turbulent, small vortices would occasionally arise in the circulating fluid. If they rotated in sympathy with the shear flow, like the Great Red Spot, they would persist for some time; if they rotated against the shear, they would be pulled apart. In a flow containing pre-existing, large rotating vortices, a vortex rotating with the flow persisted whereas one rotating against it was rapidly stretched and pulled apart. The persistent vortex then proceeded to feed on smaller vortices with the same sense of rotation that arose subsequently in the turbulent flow (Plate 4a). If two large vortices with the 'right' rotation were set up in the initial flow, they would rapidly merge into one (Plate 4b).

These calculations suggested that, once formed, a single large vortex is the most stable structure in this kind of flow. But how might it get there in the first place? Inspired by Marcus's calculations, Joel Sommeria, Steven Meyers, and Harry Swinney from the University of Texas at Austin devised experiments to investigate this kind of flow in the flesh, as it were. They used a rotating annular tank into which they pumped water at various points in the tank's base equally spaced from the centre. Outlet ports located in the base of the tank allowed the fluid to escape again. By using this pumping system rather than just filling a plain tank with water, the interaction between flow induced by pumping and extraction at different radii and flow induced by rotation of the tank set up counter-rotating zonal jets like those on Jupiter.

The researchers found that stable vortices appeared in the tank at the boundaries of the zonal jets. The vortices were situated at the corners of regular polygons: a pentagon for five vortices, a square for four and a triangle for three. The number of vortices decreased as the shearing (which depended on the pumping rate) got stronger; eventually only a single large vortex could form (Fig. 2.15). Arising spontaneously from small random fluctuations in the turbulent flow, this vortex then remained stable and more or less isolated from the rest of the flow. Dye injected into it was trapped (Plate 5); dye injected outside stayed excluded. Occasionally other small vortices, rotating in the same sense, appeared in the flow, lasting only for a short time before either merging with others or, ultimately, being swallowed up by the large vortex—just as Marcus had found.

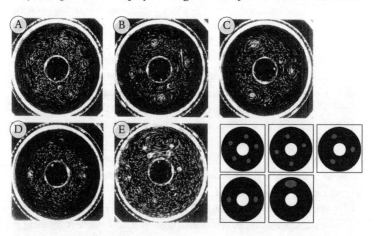

FIG. 2.15 In this experimental simulation of Jupiter's atmospheric flow, a fluid is pumped into a rotating tank so as to mimic the zonal-jet structure. Organized vortices arise spontaneously and persist in the flow. As the shear flow gets stronger, the number of vortices decreases from five (a) to one (e). The positions of the vortices are shown schematically in the images on the lower right, for clarity. (Photos: Harry Swinney, University of Texas at Austin.)

This same process has been seen on Jupiter itself: as they passed the planet in the early 1980s, the Voyager 1 and 2 spacecraft repeatedly saw small white spots approach the Great Red Spot from the east and become trapped 'in orbit' around the spot's edge before finally merging with it (Fig. 2.16). We have good reason to think, then, that Jupiter's bleary eye is a fundamental feature of its turbulent skies. Even if the present spot dissipates, another can be expected to emerge.

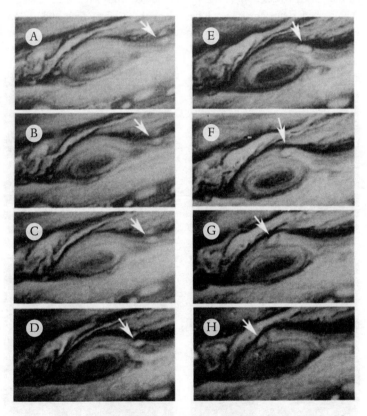

FIG. 2.16 The Great Red Spot consumes smaller vortices created in Jupiter's shear flow. In this sequence of images, taken over a period of about two weeks, a small spot (indicated with a white arrow) enters in the upper right corner and is dragged into orbit around the Great Red Spot until eventually being sucked in. (Photos: NASA.)

MANY SIDES TO THE VORTEX

Not all whirlpools are round: some are triangular, square, hexagonal, or shaped like other regular polygons. This surprising discovery was made in 1990 by Georgios Vatistas, working at Concordia University in Montréal, Canada. Vatistas set a layer of water rotating in a cylindrical tank by spinning a disk on the bottom of the vessel. As the disk got faster, the core of the vortex whisked up in the water changed from circular to having a many-lobed shape: first two lobes, then three, then four and so on (Fig. 2.17). In effect, this is equivalent to a change from a smooth, circular

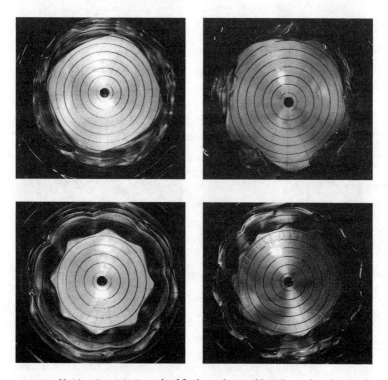

FIG. 2.17 Vortices in a spinning tub of fluid may be roughly polygonal, with several 'corners'. (Photos: Georgios Vatistas, Concordia University.)

vortex wall to a wavy one, with increasing numbers of waves fitting around the circumference, their peaks being the 'corners'. Lord Kelvin first proposed in the nineteenth century that vortex walls could develop these wavy instabilities. Vatistas thinks that, since the rotating clouds of gas and dust in spiral galaxies are comparable to vortices in fluids, the existence of these many-lobed vortex cores might explain why some galaxies seem to have not just one but several dense cores: the Andromeda galaxy, for example, has a double core, while others have several.

Curiously, an analogous patterning process happens 'in reverse' for spinning droplets of liquid suspended in a void. The Belgian physicist Joseph Antoine Ferdinand Plateau, whose experiments on soap films we encountered in Book I, discovered in the 1860s that a spinning droplet becomes deformed into a two-lobed 'peanut' shape when it rotates fast enough. Plateau too was interested in astrophysical implications—he wondered whether his droplet might mimic rapidly spinning stars or planets. Richard Hill and Laurence Eaves of the University of Nottingham in England have performed a more sophisticated version of Plateau's experiment (which used droplets of oil suspended in a water-alcohol mixture) by using strong magnetic fields to levitate droplets of water as big as grapes (14 mm in diameter). They spun these drops by passing an electric current through them to create a sort of 'liquid motor'. Hill and Eaves saw the droplets develop three, four and five lobes—crudely speaking, becoming triangular, square and pentagonal—as the spinning got faster. It seems possible that some fast-spinning asteroid-like objects in the so-called Kuiper belt beyond the orbit of Pluto, some of which are piles of rubble held together loosely by gravity, might also have three lobes. For both vortices and droplets, then, rotation can *break the symmetry*, transforming an initially circular object into one with 'corners'.

Polygonal vortices do seem to exist in nature. The eyes of hurricanes have been sometimes seen to be many-sided, with shapes ranging from triangular to hexagonal. Hurricane Ivan, which ravaged Grenada and Jamaica in 2004, had a roughly square eyewall as it approached the US east coast (Fig. 2.18). And the north pole of Saturn is surrounded by a remarkable hexagonal structure in the giant planet's atmosphere, which

FIG. 2.18 The 'square' eyewall of Hurricane Ivan.

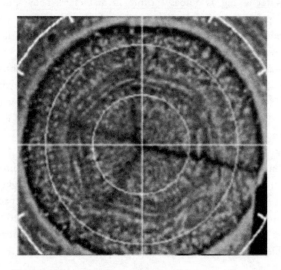

FIG. 2.19 The hexagon at Saturn's north pole. This is a persistent feature of the atmospheric flow here, but remains unexplained. (Photo: NASA.)

was discovered in the 1980s by the Voyager spacecraft (Fig. 2.19). Could these be the same structures as those we can see in a bucket of swirling water? This is not clear: the comparison only really holds if the flows have similar Reynolds numbers, whereas those on a planetary scale tend to have rather larger values of this quantity than those in the lab experiments. We have yet to fully understand why Saturn is hexed.

ON A ROLL

How Convection Shapes the World

O ne of the classic experiments in pattern formation was first described formally in 1900 by the French physician Henri Bénard, but people had surely been conducting it inadvertently for centuries in the kitchen. As I explained at the beginning of Book I, if you gently heat oil in a shallow pan, it starts to circulate in roughly hexagonal cells (Fig. 3.1). These may be revealed by adding powdered metal to the pot, so that the flakes glint as they rise and fall with the flow (though not, of course, if you have culinary aims in mind).

D'Arcy Thompson was delighted with Bénard's discovery, although he tells us that the German doctor Heinrich Quincke had seen these *tourbillons cellulaires* (cellular swirls) long before. He says that

> The liquid is under peculiar conditions of instability, for the least fortuitous excess of heat here or there would suffice to start a current, and we should expect the system to be highly unstable and unsymmetrical...[but] whether we start with a liquid in motion or at rest, symmetry and uniformity are ultimately attained. The cells draw towards uniformity, but four, five or seven-sided cells are still to be found among the prevailing hexagons...In the final stage the cells are hexagonal prisms of definite dimensions, which depend on temperature and on the nature and thickness of the liquid

FIG. 3.1 When heated uniformly from below, a layer of fluid develops convection cells in which warm, less dense fluid rises and cool, denser fluid sinks. (Photo: Manuel Velarde, Universidad Complutense, Madrid.)

layer; molecular forces have not only given us a definite cellular pattern, but also a 'fixed cell-size' ... When bright glittering particles are used for the suspension (such as graphite or butterfly scales) beautiful optical effects are obtained, deep shadows marking the outlines and the centres of the cells.

This is not only an elegant description but a perceptive one. Thompson points out that where we might expect turbulent chaos, we get geometric order, and moreover that there is a selection process that determines the size of the pattern features. It set him thinking about hexagonal patterns in layers of living cells, in soap froths, in the pores that perforate the shells of marine micro-organisms, and in the cloud patterns of a 'dappled or mackerel sky'.

To understand Bénard's observation, we need first to know what Thompson meant when he spoke about the 'peculiar conditions of in-stability' that the heated liquid experiences. The least amount of excess heat in any part of the liquid layer, Thompson says, will start a circulating current—because of *convection*.

A fluid is generally less dense when warm than when cool.* Its molecules are all jiggling with thermal energy, and the hotter they are the more they jiggle. This means that each takes up more space, so the warmed fluid expands and becomes less dense.

Now let's think about what this implies for a pan of fluid heated from below. The lower layer of fluid becomes warmer and less dense than that above it. This means it is more buoyant: like a bubble, it will have a tendency to rise. By the same token, the cooler, denser fluid on top will tend to sink. This imbalance in density is the origin of convection currents, like those that carry dust aloft above radiators in a heated room. The dust traces out the otherwise invisible motions of the air.

But if *all* the lower layer in our pan has the same buoyancy, while *all* the fluid at the top has the same ponderousness, how can they change places? Clearly the two layers cannot merely pass through one another. The uniformity—the *symmetry*—of the system hinders convection from getting under way. The only solution is to break this symmetry.

What Bénard saw was that the uniform fluid breaks up into cells in which the liquid circulates from top to bottom and back again. Bénard's cells were polygonal, but if the heating rate at the base of the pan is only just sufficient to get convection started, the cells are instead usually sausage-shaped rolls (Fig. 3.2). Seen from above, these give the fluid a striped appearance (Fig. 3.3a). Neighbouring roll cells circulate in opposite directions, so that the fluid at alternate boundaries is sinking and rising. The symmetry of the fluid is broken when these cells appear. Before that, every point at the same depth in the fluid was the same as any other. But when convection starts, a microscopic swimmer would find himself in a different situation in different locations: either buoyed up by liquid rising from below, carried along by the flow at the top of a cell, or dragged down by sinking liquid. And as D'Arcy Thompson perceived, this roll pattern has a characteristic size: the cells are about as wide as the fluid is deep.

*This is almost always the case, but, perversely, water provides an exception. It is one of the characteristic oddities of water that it is densest not when it is coldest—at the freezing point—but at four degrees centigrade above freezing. But if we are talking about heating up water from room temperature, this quirk is irrelevant: above 4 °C, water behaves 'normally', becoming less dense the warmer it gets.

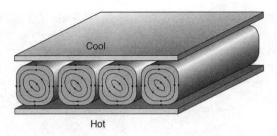

FIG. 3.2 Convection roll cells, which form in a fluid confined between a hot bottom plate and a cooler top plate. The cells are roughly square in cross-section, and adjacent cells rotate in opposite directions.

In 1916 Lord Rayleigh asked what triggered the sudden appearance of this convection pattern. It does not arise as soon as the bottom water is warmer than that at the top, even though this creates the imbalance in densities. Rather, the roll cells appear only above a certain threshold in the temperature difference between top and bottom. This threshold depends on the nature of the fluid—for example, how viscous it is and how rapidly its density changes with temperature—and also on the fluid's depth. That sounds discouraging, because it seems to imply that, if we are seeking to understand why convection starts, the answer depends on the precise details of our experiment.

But Rayleigh showed that the various factors that determine the critical threshold for convection can be combined in a single quantity that supplies a universal criterion for whether or not convection occurs. Like the Reynolds number, this parameter, now called the Rayleigh number, Ra, has no units: it is another dimensionless variable of fluid dynamics. And like the Reynolds number, the Rayleigh number specifies a ratio of forces—specifically, of the forces that promote convection (the buoyancy of the fluid, which is determined in part by the temperature difference between the top and bottom) and those that oppose it (the frictional forces that arise from the fluid's viscosity, and the fluid's ability to conduct heat and thus to even out the temperature imbalance without flowing at all). The reason convection does not arise as soon as the bottom becomes warmer than the top is because the fluid motion is opposed by friction.

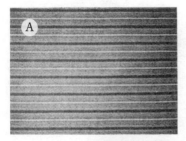
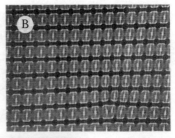

FIG. 3.3 The complexity of convection patterns increases as the driving force—the temperature difference between the top and the bottom of the vessel, measured as a quantity called the Rayleigh number—increases. First there are simple roll cells (a). At higher Rayleigh numbers, roll cells develop in the perpendicular direction too, so that the pattern consists of roughly square cells (b). At still higher Rayleigh numbers, the pattern becomes irregular (turbulent) and changes over time (c). (From Tritton, 1988.)

Only when the driving force (the temperature difference) becomes big enough to overcome this resistance do the convection cells appear. This corresponds to a Rayleigh number of 1,708.

Just as we saw when using the Reynolds number to characterize fluid flow, the beauty of treating the problem of convection by reference to the Rayleigh number is that this number is then all we need worry about (well, nearly all, as we shall see). Two different fluids in vessels of different sizes and shapes will convect (or not) in the same way when their Rayleigh number is the same. This means that one can map out the generic behaviour of convecting fluids as a function of their Rayleigh number, without having to worry about whether the fluid is water, oil, or glycerine. Rayleigh also showed that the roll cells that appear at the onset of convection have a particular width that is very nearly (but not exactly) equal to the depth of the fluid, so the rolls have an approximately square cross-section.

If the Rayleigh number is increased beyond its critical value of 1,708 to a value of several tens of thousands, the convection pattern can switch abruptly to one in which there are essentially two sets of perpendicular rolls (Fig. 3.3b). At still higher values of Ra, the roll pattern breaks down altogether and the cells take on a random polygonal appearance called a spoke pattern (Fig. 3.3c). Unlike the rolls, this pattern is not steady: the cells continually change shape over time. It is, in fact, a turbulent form of convection.

The theory of fluid dynamics, which I shall outline in Chapter 6, supplies equations for describing flow that are extremely hard to solve unless you make some simplifying assumptions. Rayleigh did just that for his analysis of convection. He considered a fluid trapped between two parallel plates and filling the gap entirely, so there is no free surface (as I showed in Fig. 3.2). Convection that takes place under this circumstance is now called Rayleigh–Bénard convection. Rayleigh also assumed that (among other things) only the density of the fluid changes with temperature—all its other properties stay the same. We know that for most fluids this is not true: they get less viscous and more runny when they are heated, for instance. And most importantly of all, Rayleigh assumed that the temperature gradient—the way in which the temperature changes from bottom to top of the fluid layer—stays constant and uniform throughout. But a rising blob of hot fluid carries heat up with it, and a cool sinking blob can cool down the lower regions. In other words, the motion of the fluid alters the very force driving that motion (the temperature gradient). Rayleigh couldn't find an easy way to take this into account.

Rayleigh found that, as Ra is increased beyond the critical value (that is, as the heating becomes fiercer), there is no longer a uniquely stable shape for the convection cells: rolls may appear that are either wider or thinner than those at the threshold itself. Physicists call these different patterns 'modes'—they are rather like the different acoustic oscillations that can be excited in an organ pipe or a saxophone's horn. Typically, the harder you blow into a saxophone, the more acoustic modes become excited and the more harmonically rich the note becomes. Rayleigh's treatment of convection shows how to calculate the range of modes that may be excited for a particular value of Ra.

In view of all its assumptions, Rayleigh's theory is surprisingly effective. It predicts correctly not only under what conditions convection starts, but also what the maximum and minimum size of the convection cells is. But within those bounds it cannot tell us anything about the shape of the cells; in fact, it cannot even show they will be roll-shaped. Moreover, to know whether a particular convection mode is truly stable, one also needs to know if all imaginable disturbances (a snake-like 'shudder' of the roll cells, say) will die out or grow bigger. Working out the stability of the various modes in the face of all such disturbances is no mean task, involving mathematical analysis considerably more complicated than that employed by Rayleigh. During the 1960s and 1970s the German physicist Friedrich Busse and his colleagues performed these difficult calculations. They discovered all manner of instabilities that might destroy the parallel sets of rolls. Busse gave these instabilities descriptive names, such as zig-zag, skewed varicose, and knot. They hedge in the options for roll cells, constraining much more tightly the permitted size and the values of Ra for which they may occur.

In fact, straight roll cells are the exception rather than the rule in experiments on Rayleigh–Bénard convection. Generally they are found only in long, narrow trays of fluid. Even here the rolls can become deformed, and strange things happen at the ends (Fig. 3.4). These edge effects can have a profound influence on the patterns in the rest of the system, which makes it harder for theorists to predict how a convecting fluid will behave and introduces a whole palette of new patterns.

In a circular vessel, parallel rolls are occasionally observed (Fig. 3.5a) but often these become distorted into a pattern that resembles the old Pan Am logo (Fig. 3.5b). This is because rolls are usually more stable when they meet a boundary wall at right angles, so the rolls bend at their ends to try to satisfy that condition. Another option is for the rolls to adapt themselves to the shape of their environment: by curling up into concentric circles, they can avoid having to meet any boundaries at all (Fig. 3.5c).

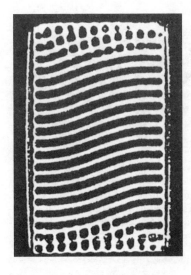

FIG. 3.4 Convection cells in a rectangular vessel. Parallel roll cells are commonly distorted by the vessel's edges. Here the rolls have a slight wavy undulation, and at the ends of the vessel they break up into square cells. (From Cross and Hohenberg, 1993, after LeGal, 1986.)

Rolls may also break up into polygonal cells, which can be regarded as a combination of two or more intersecting roll arrays. Square, triangular, and hexagonal patterns (Fig. 3.6) have all been observed; the latter are particularly common. All these patterns are predicted by Busse's complicated calculations.

Because of this rich diversity of patterns available to the convecting fluid, it is not easy to predict which will be produced in any given experiment. When several alternative patterns are possible in principle for a particular set of conditions, which is selected may depend on how the system is prepared—that is, on the *initial conditions* and the way in which these are changed to reach a specific set of experimental parameters. Pattern formation is then dependent on the *past history* of the system.

Some convection patterns also change over time. In cylindrical dishes, the regular patterns described above are unusual; more often the convection cells form an irregular network of worm-like stripes which shift

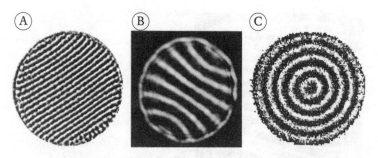

FIG. 3.5 In a circular dish, roll cells take a variety of shapes. They may remain parallel (*a*), or curve gently into a pattern resembling the old Pan-Am logo (*b*) to reduce the angle at which the rolls meet the wall. There are no such intersections at all if the cells take on the form of concentric circles (*c*). In (*a*) the fluid is carbon dioxide gas, in (*b*) it is argon gas, and in (*c*) water. (Images: *a* and *c*, David Cannell, University of California at Santa Barbara; *b* from Cross and Hohenberg, 1993, after Croquette, 1989.)

position constantly (Fig. 3.7*a*), like a mutable fingerprint. Although these patterns are disordered, nonetheless they clearly retain some vestiges of a pattern with identifiable features, such as the way all of the wavy rolls tend to intersect the boundaries more or less at right angles. One way of looking at this pattern is to see it as a set of parallel rolls disturbed by lots of 'defects' where the rolls are misaligned or broken. These defects can be classified into several types (Fig. 3.7*b*—you should be able to spot all of these in *a*). All of them are familiar from the physics of crystals, where analogous flaws crop up in the alignments of rows of atoms. Such defects can also be found in liquid crystals, in which rod-shaped molecules become aligned with each other like floating logs (Fig. 3.8). They are seen, too, in the patterns formed by the buckling of skin that give rise to real fingerprints (see Book I, page 25). While some of the cores of these curving patterns consist of concentric rolls like those in Fig. 3.5*c*, others are spirals. Convection spirals may consist of a single coiled roll cell or of two or more intertwined coils (Fig. 3.9; here you can see the double-coiled structure by looking at the centre). Researchers at the University of California in Santa Barbara have seen such spirals with up to 13 arms.

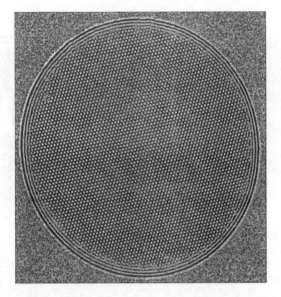

FIG. 3.6 Crossed roll cells may generate square or hexagonal patterns, as seen here in convecting carbon dioxide gas. (Notice that there are still two circular rolls running around the perimeter.) (Image: David Cannell, University of California at Santa Barbara.)

SURFACE MATTERS

Henri Bénard saw not stripes but polygons in a convecting liquid. Hexagonal patterns do appear in Rayleigh–Bénard convection, where they can be regarded as the intersection of three roll-like patterns. But Bénard himself did not study Rayleigh–Bénard convection in the strict sense, because Rayleigh's theory applies to a fluid filling the space between two plates whereas Bénard's fluid was a shallow layer with a free surface exposed to air. This surface has a surface tension, the influence of which may dominate pattern formation.

The surface tension of a liquid changes with temperature: typically, the cooler the liquid, the larger its surface tension. If the temperature of a liquid surface varies from place to place, the stronger surface tension in the cooler regions pulls warmer liquid towards it—in other words, the liquid flows across the surface from hot to cold. Upwelling of hot fluid due to

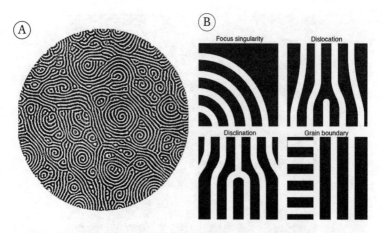

FIG. 3.7 Convection rolls may become twisted and fragmented into disordered patterns that constantly change over time (a). Several characteristic types of 'defect' can be identified in these patterns (b). (Image: a, David Cannell, University of California at Santa Barbara.)

buoyancy-driven convection can set up temperature differences at the surface: the fluid is hotter over the centre of a rising plume than it is all around. If the resulting imbalance in surface tension is the same in all directions around the plume's centre, there can be no surface-tension-driven flow because the forces pull equally in all directions. But any tiny, chance disturbance of this horizontal balance of surface tensions triggers a symmetry-breaking transition that leads to surface flow. As the fluid is pulled laterally across the surface to regions of higher surface tension, more fluid is pulled up from below to replace it. So, again, there is an overturning circulation; but now it is driven by surface tension rather than by buoyancy.

Fluid flows induced by surface-tension differences were studied in the nineteenth century by the Italian physicist Carlo Marangoni, and they now carry his name. Whether or not a flow will be created by such a difference depends on the balance between the pull of the surface tension and the resisting influences of viscous drag and heat diffusion (which neutralizes the surface-tension difference). And so there is a critical threshold for Marangoni convection, determined by a dimensionless quantity called the Marangoni number, a measure of the ratio of these opposing forces.

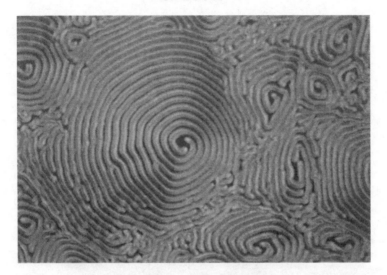

FIG. 3.8 Dislocations and other defects can be seen in the patterns formed by liquid crystals. Here, each 'cell' corresponds to a region of the liquid in which the rod-like molecules are aligned with one another in different directions. These differences can be revealed by illuminating the material with polarized light. Each sinuous domain here is just a few thousandths of a millimetre wide. (Photo: Michel Mitov, CEMES, Toulouse.)

Convection in Bénard's experiment is dominated by the Marangoni effect, which sustains the flow and determines the pattern of the convective cells. This means that the onset of convection cannot be predicted in this case by Rayleigh's theory. What's more, the most stable pattern consists not of roll cells but of hexagonal cells, in which warm fluid rises in the centre, is pulled outwards over the surface by the Marangoni effect, and sinks again at the hexagon's edges (Fig. 3.10). The differences in surface tension pucker the liquid surface, counterintuitively depressing it in the middle of the cells (where the fluid is rising) and raising it at their edges (where the fluid sinks).

REARRANGING THE ELEMENTS

D'Arcy Thompson suspected that convection patterns might explain the way skies become dappled with clouds. He was right, for the atmosphere

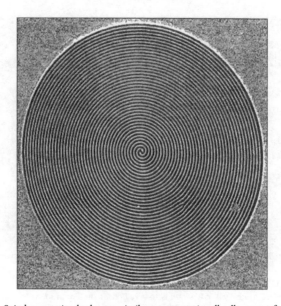

FIG. 3.9 Spiral convection looks very similar to concentric roll cells, except for a defect at the pattern's centre where the distinct spiral cells meet. Notice that the spirals also contain other defects—two are evident around the middle of the pattern in the lower left and right. The spiral structures are not stationary, but rotate slowly. (Image: David Cannell, University of California at Santa Barbara.)

is for ever churned by convection currents, and clouds are their progeny. The atmosphere loses heat by radiation from the upper layers, while sunlight absorbed by the ground and radiated as heat warms up the lower layers. This warm air rises, and often carries with it water vapour evaporated from the Earth's surface. As the air cools, the vapour condenses into droplets that reflect sunlight, creating a dense white blanket that becomes puffed into billows, or spreads in a sheet, or gathers into all sorts of strange shapes that people have mistaken for portents or UFOs.

When atmospheric circulation becomes spontaneously patterned, the clouds follow suit. Where warm, moist air rises at the edge of a convection cell, water vapour condenses at these boundaries, while dry, cold air sinks in the middle. The result is an array of cells traced out in a web of clouds, with clear sky at their centres (Fig. 3.11a). If the circulation happens in the opposite sense, the warm air rising in a central plume that diverges at the

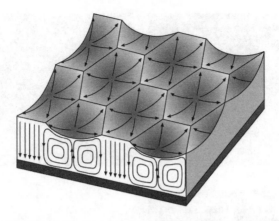

FIG. 3.10 Marangoni convection takes place in liquids with a free surface. Although it gives rise to hexagonal cells like those that can be seen in Rayleigh–Bénard convection, the origin of the pattern is different. It results from imbalances in surface tension owing to variations in temperature at the liquid surface. This makes the liquid surface pucker up as liquid is pulled from the centre to the edges of the cells.

top, then the cells have cloudy centres separated by a network of open edges (Fig. 3.11*b*). Alternatively, the convection cells might be roll-shaped, giving rise to parallel rows called cloud streets (Fig. 3.11*c*). Rayleigh's theory of convection cannot accurately describe these atmospheric motions, because some of the assumptions he made about the behaviour of the fluid are violated rather strongly by air. The roll cells that produce cloud streets, for example, are typically much wider than they are deep, unlike the roughly square-profiled rolls of Rayleigh–Bénard convection.

On much larger scales, vast atmospheric convection cells are set up by the differences in temperature between the tropics and the polar regions. These cells don't have a simple, constant structure, and moreover they are distorted by the Earth's rotation. Nevertheless, they do create characteristic circulation features such as the tropical trade winds and the prevailing westerly winds of temperature latitudes. The English astronomer Edmund Halley first proposed in the seventeenth century that convection owing to tropical heating drives atmospheric circulation, and for some time afterwards scientists believed that a single convection cell in each

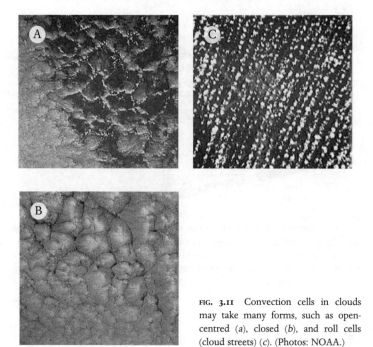

FIG. 3.11 Convection cells in clouds may take many forms, such as open-centred (a), closed (b), and roll cells (cloud streets) (c). (Photos: NOAA.)

hemisphere carried warm air aloft in the tropics and bore it to the poles, where it cooled and sank. We now know that is not the case. There are in fact three identifiable cells in the mean hemispheric circulation of the lower atmosphere: the Hadley cell, which circulates between the equator and a latitude of about 30°; the Ferrel cell, which rotates in the opposite direction at mid-latitudes; and the polar cell, which rotates in the same sense at the pole (Fig. 3.12). The polar and Ferrel cells are both weaker than the Hadley cell and are not clearly defined throughout all the seasons. Where the northern Hadley and Ferrel cells meet, the effect of the Earth's rotation drives the strong westerly jet stream.

The oceans are also stirred by convection patterns. Like the atmosphere, they are warmed in the tropics and cooled in the polar regions. This helps to establish a vast conveyor-belt circulation from the tropics to high latitudes, and the warm water carried polewards in the Gulf Stream at

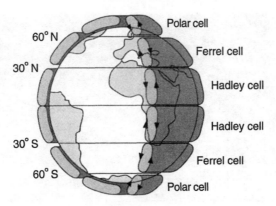

FIG. 3.12 Large-scale convection in the Earth's atmosphere is organized into three roll-like cells in each hemisphere: the Hadley cell between the equator and about 30° latitude, the Ferrel cell at mid-latitudes, and the polar cells.

the top of the North Atlantic convection cell brings with it heat that keeps northern Europe temperate (Fig. 3.13). This circulation pattern is not purely thermal (heat-driven), however. The density of sea water is also determined by the amount of dissolved salt it contains: the more saline the water, the denser it is. The salinity depends on evaporation, which removes water vapour and leaves behind saltier water, and also on freezing, since ice doesn't accommodate much salt. Thus the large-scale pattern of ocean convection is influenced by evaporation in the tropics and freezing at the poles: together, these processes give rise to the so-called ocean thermohaline ('heat-salt') circulation that regulates the Earth's climate.

As well as elaborating the sky and sea, convection shapes the slow rock of the solid Earth. Our planet is a vast convecting vessel filled with a fluid that is hotter at the bottom than the top. Yes, it really is a fluid: the rocky mantle between the crust and the core is hot enough to flow like a very sluggish liquid. The planet's molten core creates temperature of almost 4,000 °C at the mantle's base, nearly 3,000 kilometres beneath our feet, while the top of the mantle (varying in depth from a hundred to just over ten kilometres) has a temperature of several hundred degrees. In addition, the mantle contains many radioactive substances that are gradually decaying and releasing their nuclear energy, heating the fluid mantle from

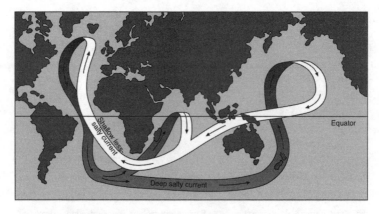

Shallow, less salty current

Equator

Deep salty current

FIG. 3.13 Convection in the oceans, driven by differences in water temperature and saltiness, creates a global pattern of conveyor-belt circulation that carries warm, less salty water along the upper belt and cold, more salty water along the lower one.

within. Even though the mantle is extremely viscous, it has a Rayleigh number of several tens of millions, and so is convecting turbulently: there are no well-ordered roll-like convection cells here, and the pattern shifts over geological time. This is what makes geophysics so interesting; indeed, you could say that it is what gives our planet a geological history, in the sense that the map of its surface is always shifting. The tectonic plates of the Earth's hard outer shell are carried along at the tops of the convection cells, and their changing positions trace out the past motions of the deep Earth. Mantle convection continuously rearranges the mosaic of the continents, ripping open new seas and instigating cataclysmic collisions. When, for example, a new surge of hot, upwelling mantle appears in the middle of a plate, as it is doing in modern East Africa, the crust is pulled apart and great rift valleys form in the divide. Elsewhere plates are pushed together, and mountain ranges like the Himalayas are created by the buckling of the crust. Some plate convergences may force one to plunge down beneath the other in a process called subduction, and the groans and judders of the sinking plate are felt at the surface as earthquakes. The tectonic plates are not passive in all this: their presence at the top of the overturning convection cells may influence the shape and disposition of the motions beneath.

The convection patterns in the mantle are further complicated by the fact that the Earth is not a set of parallel plates or a cylindrical dish, but a sphere. The patterns of convecting fluids within a sphere are not well studied for low Rayleigh numbers, let alone for turbulent motions. What is more, we do not know for sure how the mantle is structured. Some of the seismic shock waves released by earthquakes come bouncing back towards the surface from a boundary at a depth of 660–670 kilometres, which appears to split the mantle into two concentric shells. Most geologists believe that at this boundary there is a change in the crystal structure of the mantle material, brought about by the intense pressures and temperatures at these depths. Do the convection cells punch their way straight through this boundary, or do cells circulate independently in the upper and lower mantle?

Exploring these literally deep questions experimentally is complicated and relies on a lot of indirect inference. Much of what we believe about mantle convection has therefore been deduced from computer simulations. These model the mantle as a grid of tiny compartments; even if the overall flow pattern is complex, the flow in an individual compartment can be assumed to be relatively simple and may be calculated quite easily. What the simulation produces depends on what assumptions go into it: whether convection is layered or not, whether some material can pass between the layers, how much internal heat is supplied by radioactive decay, whether you include rigid tectonic plates on top, and so on.

One conclusion seems to be fairly general, however: the rising and sinking flows of mantle convection are not equivalent. The sinking fluid forms sheet-like structures called mantle slabs, which plunge back into the depths at subduction zones. It is tempting to regard the oceanic fissures where hot magma wells up to form new ocean crust, such as the Mid-Atlantic Ridge that cleaves the Atlantic almost from pole to pole or the East Pacific Rise off the west coast of South America, as the corresponding counterflows: sheets of buoyancy-driven upwelling. But these are not in fact intrinsic features of the convection pattern. Rather, the hot rock is here simply being drawn up passively from rather shallow depths by the movement of the crust away from the surface cracks, just as blowing across the surface of a cup of coffee pulls the lower liquid up to take the place of that which flows away horizontally. The fundamental buoyancy-driven upwelling structures of mantle convection seem instead to be plumes: cylindrical columns of hot, rising magma. These plumes meet the surface at hotspots, which are centres of volcanic activity.

Mantle plumes have been investigated experimentally by simulating the geological convection process in tanks of shallow viscous fluids such as silicone oil and glycerine. These experiments show that convection plumes have a mushroom shape (Fig. 3.14a), with a broad head and edges that twist into scroll-like spirals, capturing ('entraining') fluid within them. The plume head is like a three-dimensional version of the twinned vortices of the turbulent jets we saw earlier (Fig. 2.13). They may be seen inverted in the shape of ink drops descending through water, as depicted in *On Growth and Form* (Fig. 3.14b), and they show how again a turbulent liquid may organize itself into a robust and orderly structure. D'Arcy Thompson considered these bell-like shapes to be reflected in the forms of jellyfish and other soft marine invertebrates (Fig. 3.14c), and he speculated whether fluid flow might somehow be responsible for them.

The diameter of a mantle plume's mushroom head depends on how far it has travelled: if plumes begin close to the base of the lower mantle, as proponents of whole-mantle convection believe, the head can be around 2,000 kilometres across by the time it reaches the top of the mantle. There it might burst forth in a huge outpouring of molten rock, laying down vast 'flood plains' of basaltic rock. The basalt provinces found in some parts of the world, such as the Deccan Traps in western India, a region covering half a million square km and formed from more than half a million cubic km of molten rock, might bear testament to the surfacing of a deep mantle plume. Plumes that rise from shallower depths have much smaller heads when they surface as hot spots. As the tectonic plates pass across oceanic hot spots, the episodic discharge of blobs of magma creates chains of islands, such as those of the Hawaiian group.

Why are the rising and sinking features of mantle convection so different? Partly this may be what comes of having an internal heat source (radioactive decay) in the fluid. But the question also hinges on whether the mantle convects as a whole or in layers: some computer simulations have suggested that upwelling and downwelling flows are similar, and roll-like, for a separately convecting upper mantle, whereas downwelling sheets are formed if the mantle convects as a whole. The question is still

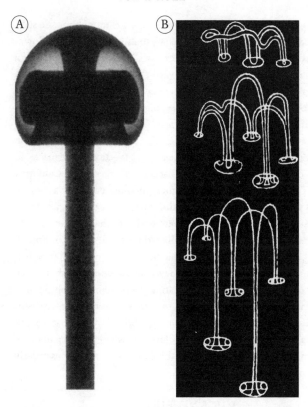

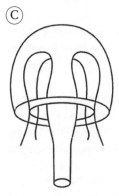

FIG. 3.14 Convection in viscous fluids at high Rayleigh number creates mushroom-shaped rising plumes (*a*). Such features are thought to exist in the Earth's mantle. Where a plume breaks through the crust, there is volcanic activity. D'Arcy Thompson recognized this same form inverted in the descent of ink droplets in water (*b*), and he also found it echoed in the shapes of some jellyfish, such as *Syncoryme* (*c*). (Photo: *a*, Ross Griffiths, Australian National University, Canberra.)

unresolved, and the evidence is conflicting. It looks as though mantle slabs do penetrate through the 660-km boundary, uniting the upper and lower mantle. But the boundary isn't invisible to descending slabs: some seem to get deflected there from their downward course, as though they hit a wall that can't easily be breached. Moreover, the chemical composition of igneous rocks at the Earth's surface seems to require that some parts of the mantle have been persistently isolated from others, whereas whole-mantle convection should stir all the ingredients together. The common view emerging now is that both styles of convection take place. In simulations of the mantle by Paul Tackley, then at the California Institute of Technology, and his colleagues, the flow pattern was organized into hot rising plumes and cold sinking sheets. The plumes were able to force a passage from the base of the mantle right to the top. But the cold sinking sheets (the mantle slabs) generally stopped at the 660-km boundary, where the cold, dense fluid accumulated in spreading puddles. When these cold pools became large enough, they would suddenly flush through to the lower mantle in an avalanche, creating a broad sinking column that then spread in a vast pool above the core. It is now believed that this episodic arrest and penetration at the 660-km boundary can happen for rising plumes too (Fig. 3.15). One thing is for sure: whatever the pattern of convection in the deep Earth, it isn't anything like as constant or as orderly as that in Bénard's dish.

ICE AND FIRE

Convection seems to be an organizing force in geology at smaller scales, too. Its characteristic polygonal imprint may be seen petrified into stone and rock in the frozen wastes of Alaska and Norway. These remote tracts may become covered with stone circles, labyrinths, networks, islands, and stripes (Fig. 3.16), the pattern features typically a metre or so across. When the Swedish geologist and polar explorer Otto Nordenskjold came across these examples of 'patterned ground' in the early twentieth century, he proposed that they are produced by circulating flows of water in soil owing to seasonal cycles of freezing and thawing.

When the frozen ground is warmed, ice in the soil thaws from the surface downwards, so the liquid water is warmer the closer it is to the

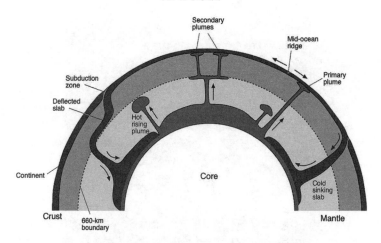

FIG. 3.15 Convection in the Earth's mantle is driven by the heat of the core, and also by the heat released as radioactive elements decay within the mantle. It tends to produce rising plumes of hot, sluggish rock, and descending slab of cooler rock. The pattern of circulation appears to be modified by a change in the chemical composition of the mantle at a depth of around 660 km, which creates a barrier (probably only partial) to the flow.

surface. For most liquids this would mean that the density simply increases with depth, which is a stable arrangement. But as I indicated earlier, water is not like other liquids: it is densest not at freezing point (0 °C) but at 4 °C. So water warmed to a few degrees above freezing close to the ground surface is denser than the colder water below it, and convection will begin through the porous soil (Fig. 3.17). Where warmer water sinks, the ice at the top of the frozen zone (the so-called thaw front) melts, while the rising of cold water in the ascending part of the convection cells will raise the thaw front. In this way, the pattern of convection becomes imprinted into the frozen zone beneath it. Such polygonal patterns can also be found on the beds of northern lakes when the water is shallow enough to freeze down into the lake bed (Fig. 3.18).

William Krantz and colleagues at the University of Colorado at Boulder have shown how this process might account for the orderly piles of stones on the surface. They say that sub-surface stones are gathered in the troughs of the corrugated thaw front, and then brought to the surface

FIG. 3.16   The freezing and thawing of water in the soils of northern tundra sets up convection currents because of the unusual way that cold water's density changes with temperature. The imprint of this circulation can be seen in polygonal cells of stones at the ground surface. Shown here are stone rings on the Broggerhalvoya peninsula in western Spitsbergen, Norway (a) and stripe-like features in the Tangle Lakes region of Alaska (b). (Photos: a, Bill Krantz, University of Colorado; b, from Kessler and Werner, 2003.)

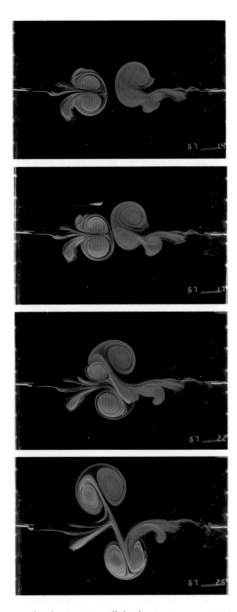

PLATE 1: When two dipolar vortices collide, the structures maintain their integrity. The mushroom-like heads exchange vortices and set off in new directions. Here they have been coloured with dyes so that they are identifiable.

(Images: GertJan van Heijst and Jan-Bert Flór, University of Utrecht)

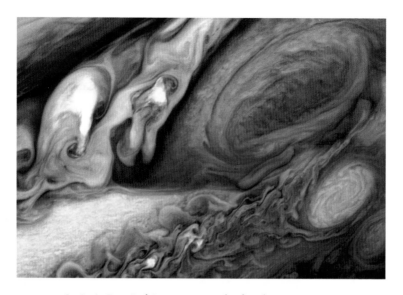

PLATE 2: Jupiter's Great Red Spot is an example of a coherent structure in turbulent flow. It has persisted in Jupiter's swirling atmosphere for at least 180 years. Other, short-lived structures have come and gone. (Photo: NASA)

PLATE 3: The bands that encircle Jupiter are flow patterns called zonal jets.

(Photo: NASA)

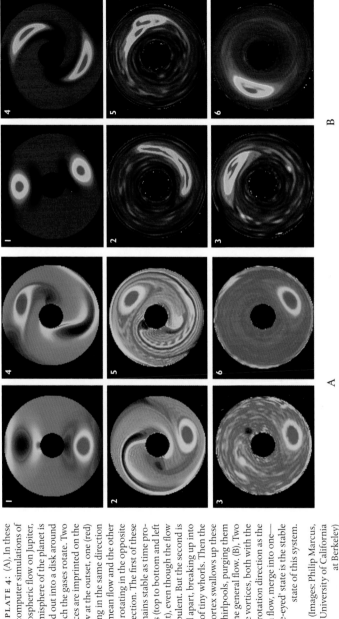

PLATE 4: (A), In these computer simulations of atmospheric flow on Jupiter, a hemisphere of the planet is flattened out into a disk around which the gases rotate. Two vortices are imprinted on the flow at the outset, one (red) rotating in the same direction as the mean flow and the other (blue) rotating in the opposite direction. The first of these remains stable as time progresses (top to bottom and left to right), even though the flow is turbulent. But the second is pulled apart, breaking up into a mass of tiny whorls. Then the red vortex swallows up these little whirlpools, purging them from the general flow. (B), Two large vortices, both with the same rotation direction as the mean flow, merge into one—the 'one-eyed' state is the stable state of this system.

(Images: Philip Marcus, University of California at Berkeley)

A

B

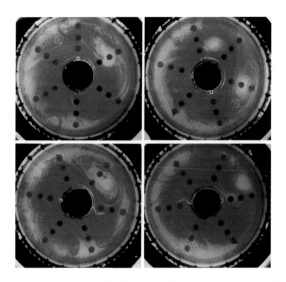

PLATE 5: An experimental model of Jupiter's flow in a rotating tank of liquid produces a model 'Red Spot', revealed here by injecting dye into the fluid. At least one vortex persists throughout, as revealed in these four snapshots

(Photo: Harry Swinney, University of Texas at Austin)

PLATE 6: Sand dunes are self-organized patterns on a grand scale.   (Photo: Rosino)

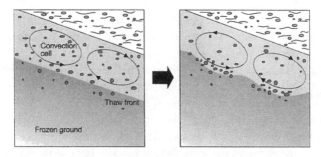

FIG. 3.17 As water circulates in convection cells through the soil, the pattern is transferred to the 'thaw front', below which the ground remains frozen. Stones gather in the troughs of the thaw front, and are brought to the surface by 'frost heaving' in the soil.

FIG. 3.18 Freeze-thaw cycles of groundwater at the edge of this Norwegian lake have produced convection cells traced out by stones on the lake bed. (Photo: Bill Krantz, University of Colorado.)

by 'frost heaving', a process familiar to farmers, that happens when soil freezes. So a field that freezes during a frost and then thaws becomes littered with stones that trace out the pattern of convection beneath the ground. Krantz and colleagues calculated the convection patterns that may arise as water circulates through porous soils. They found that polygonal (particularly hexagonal) patterns are favoured on flat ground but that the convection cells are roll-like on sloping ground, giving rise to rows of stones.

Mark Kessler and Bradley Werner of the University of California at San Diego have concocted a more detailed computer model of the process which shows how diverse the patterns can be and how they may mutate one into another. They say that as the ground freezes from the surface downwards, it does so faster where there are more stones beneath the surface than where there is just soil, because the soil holds moisture, which is slow to freeze. The net effect is for stones to be pushed not only upwards but towards regions where other stones are gathered, while soil gets pushed downwards and towards soil-rich regions. Thus, stone and soil get segregated. The stone domains also become squeezed and elongated, especially if the soil is hard to compact. Kessler and Werner's model suggests that the patterns that result depend on the concentration of stones in the soil, and also on the slope of the ground and the tendency for stone domains to be elongated: they see switches between stone holes, islands, stripes, and polygons as these factors change (Fig. 3.19). The patterns are highly reminiscent of the animal markings examined in Book I. And, curiously, polygonal

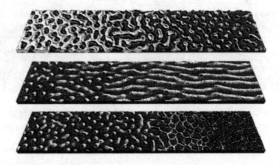

FIG. 3.19   The model of 'sorted ground' devised by Kessler and Werner generates a wide range of stone patterns. In the top image, the ratio of stones to soil decreases from left to right; in the middle, the slope of the ground increases from left to right; and in the bottom image, the tendency of the stone domains to elongate increases from left to right. (From Kessler and Werner, 2003.)

FIG. 3.20   Solar granules are highly turbulent convection cells in the Sun's photosphere. (Photo: The Swedish Vacuum Telescope, La Palma Observatory, Canary Islands.)

networks seem to follow rules for polygon-wall junctions analogous to those found in soap foams (Book I, Chapter 2): three-way intersections with equal angles of about $120°$ are preferred, and four-way junctions are unstable. These are strong hints that universal patterning rules are at play.

Convection may be seen on a grand scale on the Sun's surface. Sunlight comes from a 500-km thick layer of hydrogen gas close to the star's surface, which has a temperature of about 5,500 °C. This gas is heated from below and within, and radiates its heat outwards from the surface into space. So although it is about a thousand times less dense than the air around us, it is a convecting fluid. Its Rayleigh number is so high that we would expect it to be utterly chaotic and unstructured. But photographs of the Sun's surface show that, on the contrary, the photosphere is pock-marked with bright polygonal regions called solar granules, surrounded by darker rims (Fig. 3.20). These granules are the tops of convection cells: the bright centres are regions of upwelling and the dark edges trace out cooler, sinking fluid. They range in size from about 500 and 5,000 km across, making the largest of them comparable to the size of the Earth. The pattern is constantly changing, each cell lasting for only a few minutes, and its existence in such a turbulent fluid shows that we still have a lot to learn about the patterns of convection.

RIDDLE OF THE DUNES

When Grains Get Together

I t would be easy to suppose that Ralph Alger Bagnold, rather like T. E. Lawrence, went initially to the desert because of the army, and ended up staying in the army because of the desert. He was by all accounts a good soldier, but one cannot help thinking his mind must have been only half on the job, for he could not suppress his scientific instinct for enquiry even in the most unlikely circumstances. Trained as an engineer, he joined the British Army's Royal Engineers in 1915 and found himself posted to Egypt and India, where he fell in love with the deserts. By the 1920s he was spending his leave exploring these 'seas of sand', joining the 1929 expedition in search of the legendary city of Zerzura west of the Nile that was led by László Almásy, the Hungarian nobleman who inspired Michael Ondaatje's novel *The English Patient*.

'We forgave Bagnold everything', says Ondaatje's Almásy, 'for the way he wrote about dunes.' That is indeed what Bagnold did, and with such perception and insight that his 1941 book *The Physics of Blown Sand and Desert Dunes* became the standard work on dune formation for many decades. Inspired by his observations in Libya and informed by wind-tunnel experiments in England, Bagnold set out to explain how sand grains are organized by the desert wind into structures ranging from ripples the size of your fingers to undulations several kilometres across.

I cannot better Bagnold's description of the basic dilemma posed by the patterns of sand dunes. 'Instead of finding chaos and disorder,' he wrote, the observer never fails to be amazed at a simplicity of form, an exactitude of repetition and a geometric order unknown in nature on a scale larger than that of crystalline structure. In places vast accumulations of sand weighing millions of tons move inexorably in regular formation, over the surface of the country, growing, retaining their shape, even breeding, in a manner which, by its grotesque imitation of life, is vaguely disturbing to an imaginative mind.

What Bagnold is really saying here, though lacking the modern term, is that dunes are *self-organized*. Wind alone has no intrinsic capacity to make these stripes and crescents and other fantastic shapes, nor to set their scale. Sand ripples and dunes are a conspiracy of grains, a pattern that emerges from the interplay of windborne movement, collision-driven piling up, and slope-shaving avalanches. The formation of sand dunes is not only one of the most fertile of pattern-creating processes; it is also something of an archetype, an exemplary demonstration of how such patterns lie in wait in systems of many interacting parts, even though no amount of close inspection of the components will reveal them. We will see it is no coincidence that some of these patterns again display characteristic, seemingly universal features that we have seen before.

There is something deeply odd about grains and powders. They are made of solid stuff—sand is mostly quartz, hard and crystalline—and yet they flow. Sand supports our weight, but we can pour it from a cup. An extreme and rather terrible example of this Janus aspect can be seen during some earthquakes, such as that which shook the Marina district of San Francisco Bay in October 1989. Many of the houses there were sent tumbling as a segment of the San Andreas Fault slipped, and although miraculously there was no loss of life, hundreds of millions of dollars' worth of damage was done. Yet elsewhere in the bay area there was nothing like this degree of destruction. What brought low the Marina district is that it was built on sand-rich landfill sites. When the earth shook, these wet, sandy soils turned to a slurry that flowed like treacle. This property of a granular substance, naturally enough called liquefaction, is well known to seismologists and civil engineers. It is one of the most dramatic manifestations of the fact that a granular substance is a peculiar state of matter.

Engineers and geologists urgently need to understand such behaviour, and not only to gauge earthquake hazards. All manner of industrial substances are routinely handled in the form of granular powders, from cement to drugs to breakfast cereals, nails, nuts, and bolts. Graininess is everywhere in the geological world: it dictates the behaviour of landslides, the transport of sediments, and the shape and evolution of deserts, beaches, soils, and stony ground. There are old rules of thumb for predicting how grains behave, but only in recent years have scientists begun to appreciate that to attain more fundamental understanding they must invent new physics.

Strange things happen in grains. Shaking together different kinds of grain can mix them up, or, on the contrary, may have the opposite effect of segregating them. Sound waves can bend around corners as they travel through sand. The stress beneath a sand pile is smallest where the pile is highest. Yet the pressure at the bottom of a tall column of sand is the same regardless of the column's height, which is why a sand glass is a good timekeeper: the sand leaks out at a steady rate even though the column gets smaller.

SHIFTING SANDS

Not all deserts are sandy, and not all sands are piled into dunes; but it is the dune (Plate 6), which is found within only 20 per cent of the world's deserts, that defines our archetypal desert imagery. These seas of sand are almost barren yet intensely beautiful, both terrifying and holy. North Africans say that the desert is the Garden of Allah, scoured of life so that he may walk there in peace.

Desert dunes range in width from a few metres to several kilometres, and may themselves be organized into complex mega–dunes, sometimes called draas after their name in North Africa, which can be up to several kilometres broad. Moreover, dunes come in a variety of shapes; indeed, there are so many varieties, with region-specific names derived from the local language, that even geomorphologists have trouble keeping track of them. At the smallest scale, the desert floor is wrinkled into little wavy ridges typically about the width of your arm (Fig. 4.1). The crests of these ripples may be spaced as little as half a centimetre apart, or as much as several metres.

FIG. 4.1 Ripples in sand are self-organized patterns formed as the wind picks up and transports the grains. (Photo: .EVO.)

Bagnold sought to explain why wind-blown sand gets deposited in this wrinkled pattern. In today's terminology, we would say that what he proposed is an example of a *growth instability* driven by positive feedback, so that small disturbances get bigger.

Let's start with a flat sandy plain, across which a steady wind is blowing. The wind continually picks up grains and dumps them elsewhere. If the wind blows always in the same direction, the plain is gradually moved en masse downwind; the borders of deserts get shifted around this way. But wouldn't we expect the grains simply to be redistributed at random, so that the sand surface stays smooth?

That is what you might think; but Bagnold's growth instability makes the flatness prone to wrinkling. Imagine that, purely by chance, a little bump appears where slightly more sand has been dumped than elsewhere—that is to be expected if the scattering of grains is truly random. The windward side of the bump (called the stoss side) is now higher than the ground around it, and so it captures more sand from out of the breeze. This is illustrated in Fig. 4.2a, which shows that more lines, representing the trajectories of wind-borne grains, intersect the stoss face than they do a horizontal part of the surface with the same area. This means that the stoss slope starts to grow even taller. Conversely, fewer lines intersect the downwind (lee) side of the bump, where there is an 'impact shadow'—so the rate of grain deposition is smaller than average here, and the slope gets accentuated rather than levelled out. Once a bump is formed, therefore, it becomes self-amplifying.

That in itself seems to imply that the plain should become covered in bumps at random locations. But the ripple pattern is not random: there is a characteristic separation between ripples, a particular *wavelength*. Where does that come from? It turns out that the formation of one bump triggers the appearance of another one downwind, so that a system of ridges propagates itself across the plain. This comes about because the wind-blown grains bounce when they hit the desert floor. The wind carries these bouncing sand grains downwind in a series of hops, a process called saltation. The initial impact of a grain also creates a little granular splash, throwing out other grains from the surface which can then also be carried along by saltation.

Saltation could in itself be a smoothing-out process, since it means that the grains that hit the surface get scattered again. But when a ripple begins to form, the rate of downwind grain transport by saltation becomes uneven. On a flat surface, saltation creates a flow of jumping grains in the direction of the wind. But because there are fewer impacts on the lee slope, those that occur at the foot of this slope send grains jumping downwind (to the right in Fig. 4.2) that are not replenished by grains jumping in from the other direction (to the left here). Therefore the foot of the slope becomes excavated, and a new stoss slope develops to its right (Fig. 4.2*b*, *c*). The overall effect is that one bump spawns another just downwind, out of its impact shadow. So a single bump grows into a series of ripples. As Bagnold put it, 'a flat sand surface must become unstable, because any small chance deformation tends to become accentuated by the local sand-removing action of the saltation'.

He also suggested why the ripples have a characteristic wavelength: this is determined, he said, by the typical distance that a saltating grain travels before coming to rest (which in turn depends on the grain size, the wind speed, and the wind angle). It now seems, however, that this wavelength reflects a balance between rather more complex aspects of the grain-

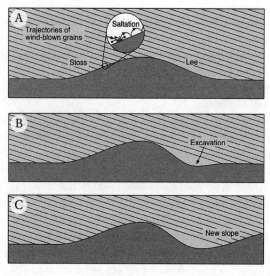

FIG. 4.2 The formation of sand ripples involves a propagating instability. Wind-borne grains rain down obliquely on the desert surface. Where the surface slopes, more grains hit the windward (stoss) side of the slope than the leeward side (*a*). Each grain scatters others from the surface as it strikes, and travels downwind for a few short hops before coming to rest—a process called saltation. The accumulation of saltating grains at the slope crest means that the leeward foot of the slope receives fewer grains than elsewhere, and so it becomes excavated into a depression (*b*). This depression develops into a new stoss slope, downwind of the first—and so a new ripple begins to form (*c*).

transport process, and in reality there is usually a range of wavelengths in any field of ripples.

There was, after all, a lot that Bagnold could not incorporate into his simple theory. With computer models it is much easier to build in some of these complexities. In such a model devised by Spencer Forrest and Peter Haff from Duke University in North Carolina, the collision of a grain with the sand surface is described by a so-called splash function, which specifies the number of grains ejected in the impact, and their velocities. As the grains are 'fired' at a flat bed of sand with a particular speed and angle, ripples quickly begin to form (Fig. 4.3a). These have a triangular cross-section, and rather than simply standing where they first appear, they move in procession across the surface in the direction of the wind, just as they do in real deserts. The ripples are not in any sense being 'blown along' by the wind; instead, their synchronized motion is an intrinsic consequence of saltation.

This motion ensured that differences in ripple size get evened out. Smaller ripples travel faster than larger ones, simply because they contain less material to be transported. But as they overtake larger ripples, small ones 'steal' sand from them until their sizes, and therefore their speeds, are more or less the same: they are several hundred times the width of individual grains (Fig. 4.3b).

In these simulations, deposition of grains meant that the sand bed gradually increased in thickness. In the real world such beds can be preserved for posterity, turned to permanent rock as the gaps between the grains are filled in with a cement of minerals precipitated from permeating water. Such sedimentary rocks are known as aeolian sandstone. (Aeolian means wind-borne, after the Greek god Aeolus, king of the winds.) By artificially colouring the wind-borne grains at regular intervals in their computer model, Forrest and Haff were able to deposit 'stained' layers which acted as markers to show how the deposited material became distributed in the thickening bed. Depending on the rate of deposition, they found various patterns (Fig. 4.3b,c), which resemble those found in natural aeolian sandstones when some environmental change (for example, a change in the chemical composition and thus the colour of the sandy material) allows material deposited at different times to be distinguished.

MARCH OF THE DUNES

Seen from above, both small-scale ripples and fields of sinuous dunes (Fig. 4.4) resemble the fingerprint-like stripes seen in convection and in some of the chemical 'activator-inhibitor' patterns I discussed in Book I. (Look in particular at where ripples terminate or bifurcate in two.) The biologist Hans Meinhardt at the Max Planck Institute in Tübingen suggests that, at root, the formation of these sand patterns is

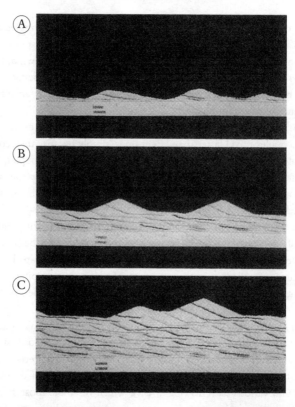

FIG. 4.3 Self-organized ripples forming in a computer model of wind-blown sand deposition. The ripples spring up from random irregularities on an initially flat surface (a). The ripples move from left to right because of saltation (the downwind jumping of grains). Small ripples move faster than large ones, and so ripples may collide and exchange grains until their size, speed and spacing are more or less uniform (b). 'Stained' grains injected at regular intervals trace out the patterns formed by distinct layers (b,c). (Images: Peter Haff, Duke University, North Carolina. Reproduced from Forrest and Haff, 1992.)

indeed akin to an activator–inhibitor system, in which short-range 'activation' (initiation) of pattern features competes with their long-ranged inhibition (suppression). The mounds of sand are formed by deposition of wind-blown grains. As a ripple or dune gets bigger, it enhances its own growth by capturing more sand from the air. But in doing so, the dune removes the sand from the wind, and also shelters the leeward ground, both of which suppress the formation of other dunes in the vicinity. The balance between these two processes establishes a constant mean distance between dunes. And yet the pattern is not static: like sand ripples, dunes are constantly on the move, shifting in a stately, writhing dance. The pattern persists while its individual components change.

Probably the most familiar types of sand dunes are those that share the same wavelike form as sand ripples, with linear, slightly wavy crests that lie perpendicular to the wind direction. These are called transverse or linear dunes (Fig. 4.4). Others form crests parallel to the prevailing wind: these are longitudinal dunes. Some dunes have several arms radiating in different directions: these are star dunes (Fig. 4.5a). Barchan dunes are

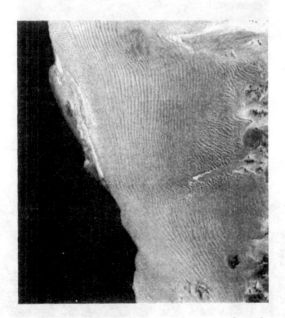

FIG. 4.4 Linear dunes in the Namib Sand Sea in south-western Africa. The area shown is about 160 km from left to right. (Photo: Nick Lancaster, Desert Research Institute, Nevada.)

crescent-shaped, with their horns pointing downwind (Fig. 4.5b). Always on the march, barchan dunes can merge into wavy crests called barchanoid ridges. The motion of barchan dunes has been tracked by a remarkable coincidence. In 1930, Ralph Bagnold took part in an expedition in northern Sudan during which the party camped one night in the lee of a barchan dune. The next morning, they left their empty cans to be buried

FIG. 4.5 Dunes have several characteristic shapes, including many-armed star dunes (a) and crescent shaped barchan dunes (b). (Photos: a, Copyright EPIC, Washington, 2003; b, Nick Lancaster, Desert Research Institute, Nevada.)

by the shifting sands. Fifty years later, this abandoned rubbish was discovered exposed on the desert floor by the American geologist Vance Haynes, who was able to confirm its identity. The great barchan dune had moved right over the pile and was now nearly 500 feet away.

These various dune types seem to be universal forms created by the interaction of wind and grains. All of them have been seen traced into the dust of Mars, along with others that are not familiar on our planet (Fig. 4.6). The question, then, is not simply how dunes form, but how the same basic grain-transport process (saltation) produces several different forms.

Many models have been proposed to account for the shapes, sizes, and arrangements of particular kinds of dune, some of them invoking rather complex interactions between the evolving dune shape and the wind-flow pattern. Bagnold suggested that longitudinal dunes might be created by helical wind vortices stirred into the wind by convective airflow above the hot desert surface. Another early pioneer of dune studies, the British geographer Vaughan Cornish, suggested at the beginning of the twentieth century that star dunes form at the centre of convection cells above the desert floor. It's clear that the nature of the wind, whether steady or varying in direction, fast or slow, has a big influence on the type of dune it generates. The amount of sand available for dune-building is also important: transverse dunes may be favoured if the sand supply is abundant, whereas longitudinal or barchan dunes form in a sparser environment. The fact that the dune itself changes the flow of air around it as it grows adds a further level of complication, as does the presence of vegetation or of complex underlying topography. So-called coppice dunes are formed when small patches of vegetation accumulate sand, while climbing dunes, echo dunes and falling dunes are caused by geographical features such as hills.

Can we, then, say anything general about the factors that create dunes? Bradley Werner has developed a computer model that generates different dune types under different conditions. He imagines grains being transported not individually but in slab-like 'parcels'. These initially lie scattered at random on a rough stony bed, and are picked up randomly by the wind. After it has been carried a fixed distance, each parcel has a particular chance of being re-deposited. The probability of that is greater if the parcel hits

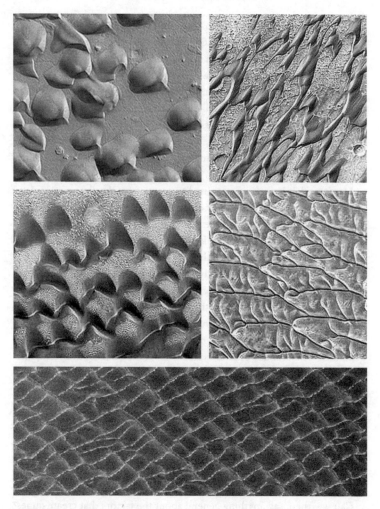

FIG. 4.6 Some of the unusual dune types seen on the surface of Mars. (Photos: NASA.)

sand-covered ground rather than stony ground, because sand bounces by saltation over stony ground more readily than over sandy surfaces—it may often 'bounce' until it finds a softer landing. If the parcel is not deposited, it is carried on for the same fixed distance before the possibility of being deposited arises again. If at any point a pile of sand gets too steeply sloping, the slabs there slide downhill until the slope is stable. This maximum allowed slope of a sand pile is called the angle of repose; we will see later that it plays a major role in the behaviour of grainy materials.

Now, Werner's model is not self-evidently a good way to describe windblown sand: it seems a little odd to parcel up the sand into irreducible slabs, for instance, or to assume that these always get carried for the same distance before 'hitting' the ground again. Some of these assumptions are understandable—dunes being so much bigger than sand ripples, it's not really feasible to simulate them grain by grain—but that doesn't mean they are reasonable. And yet the model seems to be on to something, for with just these ingredients Werner was able to reproduce all the major dune types: barchan, star, and linear dunes (Fig. 4.7). When the wind was predominantly in a single direction, dunes formed with their crests lying perpendicular to the wind (transverse dunes), whereas if the wind direction was more variable, the dunes were oriented in the average direction of the wind (longitudinal dunes). While it is likely that specific, local influences affect dune sizes and shapes, Werner's model has the appealing feature that the broad patterns that emerge are generic, not dependent on case-by-case details. Within this picture, star and barchan dunes are as inevitable a feature of nature's tapestry as the branches of a river or the stripes of a zebra.

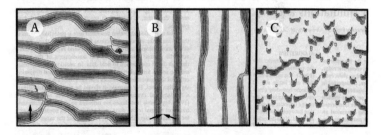

FIG. 4.7 A model of dune formation devised by Bradley Werner generates many of the common dune types, including transverse and longitudinal linear dunes (*a,b*) and barchan dunes (*c*). Here I show the contours of the deposited material. The shapes depend on the wind direction and variability, indicated by arrows. (Images: from Werner, 1995.)

Hans Herrmann at the University of Stuttgart and his colleagues are not convinced that dune formation is so simple. They say that some of the dunes that appear to emerge from models like this are merely transient structures, which will fade into a featureless bed of sand if the model is run for long enough. They think that the key factors in dune formation are actually rather subtle, tied up with the details of how the wind-borne supply of sand varies from patch to patch and how a nascent dune affects the airflow around it. Grains are not simply fired down onto the desert surface along straight-line trajectories at a fixed angle. Instead, dunes act like obstacles in a stream's flow, bending and reorganizing the streamlines. In particular, the airflow over the ridge of a dune is a little like the water flow past a bridge that we saw in Chapter 2. The streamlines bend around the obstacle, but in the wind-shadowed lee of the dune a circulating vortex may form (Fig. 4.8). Bagnold explained all this, and measurements of flow around real dunes have shown that it is indeed what happens. It means that the leeward shadow is not merely a 'dead zone' that no grains reach; the vortex here can scoop sand from the lee, eroding the slope.

A dramatic consequence of this effect of fluid flow has been seen by the physicist Atsunari Katsuki of Osaka University in Japan and his co-workers in experiments on the formation of artificial dunes. Dunes are usually too big and slow-forming to study in the lab, but Katsuki and his colleagues mimicked the process by suspending sand in flowing water, which carried the grains down a ten-metre trough. This produced barchan-shaped dunes a few centimetres in size. Just like real dunes, these miniature versions moved gradually down the trough, horns first (real barchans can move downwind at several tens of metres a year). The speed of a dune depends

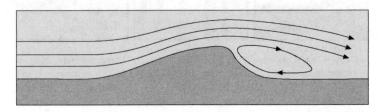

FIG. 4.8   A vortex forms behind the crest of a dune and erodes the lee slope.

on its size, smaller ones moving faster. This means that small dunes formed behind big ones can catch up with them and collide.

In some collisions, a large dune began to break up as a small one approached it from behind, even before the two came into contact (Fig. 4.9a). By the time the small barchan reached the big one, the larger dune had split in two. This strange behaviour, Katsuki and colleagues realized, was caused by a vortex of fluid stretching away from the lee of the small dune, which eats away at the stoss slope of the big one.

There were two other kinds of collision too, depending on the sizes (both relative and absolute) of the colliding dunes (Fig. 4.9b,c). In one, the dunes simply merge; but the other is perhaps the oddest of all: the small dune appears to pass straight through the big one. How is that possible, without the grains getting scrambled? According to Hans Herrmann, who has used a computer model to simulate these dune collisions, the grains

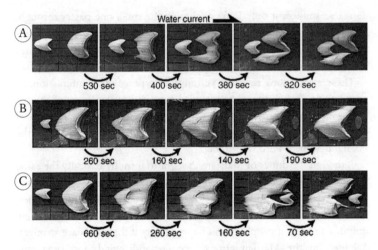

FIG. 4.9 In experiments in a water flume that mimic the formation of sand dunes, the collision of a small, fast-moving dune with a larger, slower one could have several outcomes. The large dune could be split in two as the small one approaches (a), the small dune could simply merge (b), or it could apparently pass right through (c). (Photos: Endo *et al.*, 2004. Copyright 2004 American Geophysical Union. Reproduced by permission of American Geophysical Union.)

FIG. 4.10 A computer model of dune formation suggests that another result of these dune collisions can be the spawning of two barchan dunes from the 'horns' of the large one (*a*). This might explain the clusters of dunes seen in some deserts (*b*). (Photo and image: Hans Herrmann.)

do get scrambled: it only looks as though the small one passes through. What really happens, Herrmann says, is that the small dune cannibalizes the big one: it grows and slows down, while the big dune shrinks and speeds up.

These simulations revealed yet another type of dune interaction: as they collide, 'baby' dunes are spawned from out of the horns of the big one (Fig. 4.10*a*). Something like this may be responsible for clusters of different-sized barchan dunes seen in nature, like one observed in the coastal deserts of Peru (Fig. 4.10*b*).

Although the dunes of Mars are thought to form by basically the same processes of wind-blown transport and saltation, there is an important difference on the red planet: its atmosphere is about 100 times thinner. Saltation occurs only if the friction of the wind on the grains is big enough, and in a thinner atmosphere this happens only if the winds are stronger. Saltation in the Martian atmosphere demands winds ten times the strength of those on Earth. But gales as formidable as that do occur on Mars. Because the conditions of dune formation differ in this way, so do the shapes of some of the dunes. Herrmann and his colleague Eric Parteli find that their model can reproduce some of the Martian dune shapes unknown on Earth (Fig. 4.11).

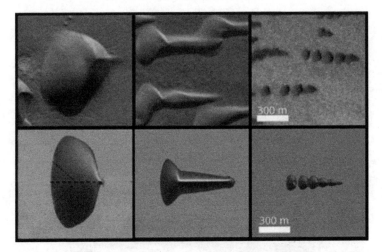

FIG. 4.11 The model can also reproduce some of the dune shapes on Mars, where the atmospheric pressure is lower but the wind speed can be higher. Here the upper frames show real Martian dunes, and the lower frames show corresponding forms generated in the computer model. (Images: from Parteli and Herrmann, 2007.)

STRIPED LANDSLIDES

One intriguing feature of natural sand patterns—both small-scale ripples and large dunes—is that the sand grains are segregated by size into different parts of the mound. In sand ripples, the coarsest grains tend to accumulate at the crests and in a thin veneer coating the stoss face. For large dunes it's often the other way round: the finest grains collect at the crests, and the coarsest in the troughs. As sand continues to rain down so that ripples are gradually laid on top of each other, the result is a series of stratified layers: a periodic alternation of coarse and fine grains down through the sand bed.

How does this grain-size sorting occur? Robert Anderson and Kirby Bunas from the University of California at Santa Cruz have shown that it can be produced by saltation. They studied a model rather like that of Forrest and Haff, except that it incorporated grains of two different sizes, each with a different splash function: large grains ejected more secondary grains, since their collisions were more energetic. The size and speed of

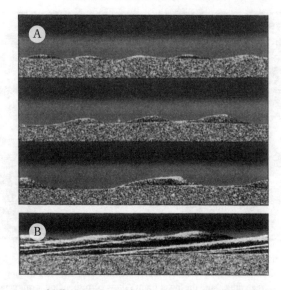

FIG. 4.12 Grains of different sizes are often segregated in sand ripples and dunes. Here a computer model reveals the tendency of sand ripples to accumulate coarse grains (white) on the stoss slopes, and particularly at their crests (a). When the deposited layer gradually thickens, the sand deposit becomes stratified (b). (Images: Robert Anderson, University of California at Santa Cruz.)

the impacting grain, as well as the composition of the bed that it struck, also determined the relative mixture of small and large grains in the 'splash'. Thus the rules governing the impacts were fairly complicated; but their net effect was that impacts tended to throw out the smaller grains preferentially, and with higher speeds (which carried them further away). The general effect of impacts was therefore to make the surface of the sand bed coarser.

Because the stoss slope receives more impacts, it gets more coarsened (Fig. 4.12a), as seen for real-world ripples. And since the larger grains make smaller hops, they gradually make their way up the stoss slope and jump just over the crest into a sheltered region at the top of the lee slope within the impact shadow. Here they remain, protected from impacts, while further coarse material gradually climbs on top of them. The smaller grains, meanwhile, make bigger leaps and so are propelled further over the edge onto the lower parts of the lee slope. As a result, the crests of the ripples are particularly enriched in coarse grains, again as seen in nature.

Notice that the ripples here are asymmetric, with a gently convex stoss slope and a steeper, concave lee slope. This shape is much closer to that of real sand ripples than are the triangular mounds of Forrest and Haff's model, showing that the more sophisticated treatment of saltation and splashes does a better job of imitating the real process. And because the ripples are not static but are slowly moving downwind, the coarse material on the ripple crests is repeatedly buried and then exhumed again at the foot of the stoss slope as the ripples pass over it: the grains are for ever climbing mountains. So as the sand bed gradually thickens, stratified beds are laid down in which coarse and fine layers alternate (Fig. 4.12*b*), mimicking the graded layers of aeolian sandstone.

There is another, quite different way of using nature's self-organizing capacity to sort grains of different types into layers. As it is both simple and reliable, I have used it several times in demonstration lectures to show how easy pattern formation can be. In fact, the process is so simple that it is astonishing it was apparently not described until 1995, when it was discovered by Hernán Makse and Gene Stanley from Boston University and their colleagues. All you do is to mix up grains of different sizes and shapes— coarsely granulated sugar and fine sand will do (the pattern is not easy to see unless they are differently coloured)—and pour them into a heap. As the heap builds up and grains tumble down its slopes, the two types of grain become segregated into layers parallel to the slope. This is seen most easily by making a two-dimensional heap (a slice of the conical mound) so that you can see the cross-section (Fig. 4.13). That can be done by pouring the mixture* into the gap between two transparent sheets of glass or perspex (Plexiglas). Once the heap grows big enough, there are regular landslides down the slopes— and you will see, I suspect with some amazement, the stripes unfurl before your eyes.

This pattern-formation seems to deny intuition: it is as if time were running backwards. We simply do not expect mixtures like this to

*The mixture should be stirred, not shaken. Shaking, as we will see later, does not necessarily guarantee good mixing of different kinds of grain.

FIG. 4.13 Two well-mixed types of grain of different sizes and shapes (here dyed different colours) will separate spontaneously into stripes when poured into the narrow space between two plates. Notice also the segregation of grains, with one type at the left-hand top of the slope and the other at the right-hand foot. (Photo: Gene Stanley, Boston University.)

segregate of their own accord. Indeed, I indicated in Book I that the second law of thermodynamics seems to forbid it: this law insists that things get *more* disorderly and jumbled up as time progresses. What's more, the grains don't just separate out—they separate into a pattern with a characteristic size scale, namely the width of the stripes. These stratified landslides have surely been taking place for centuries in industry, engineering, and agriculture, as for example when mixtures of different cereal grains or sands are poured out of a hopper. But perhaps the layers remained hidden inside the conical heaps.

If you do the experiment, you'll see that the stratification happens in a characteristic manner: each landslide generates a pair of stripes, which begin to appear first at the bottom of the slope and run back up it in a kind of kink at the sloping face. The topmost stripe of the pair contains the larger grains. Makse and colleagues supposed that the key to this sorting process is that the larger grains tumble down the slope more freely than the smaller ones, which are more easily trapped in small dips and irregularities. This same effect can be seen in rock slides, where the largest boulders crash to the bottom of the slope while the smaller ones get stuck

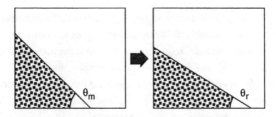

FIG. 4.14 A pile of grains will undergo an avalanche at a critical angle called the angle of maximum stability, here denoted θ_m. The avalanche 'relaxes' the slope to a stable angle called the angle of repose (θ_r). These angles generally differ for grains of different shapes.

further up the hillside. In effect, the slope looks smoother to the large grains than to the small ones.

Since the large grains reach the bottom first, there is segregation of these grains at the foot of the heap. There they pile up and create a kink in the slope. Then, as the subsequent grains tumble down and reach the kink, the small grains get stopped first, since the large ones are less easily trapped there. So the small grains are deposited first, and the large ones come to rest on top, moving the kink back up the slope as they do so.

To study this in a simple model, we need to specify some kind of criterion for when an avalanche starts. This is a well-understood issue for piles of grains, and you can see it for yourself by tipping up bowls of granulated sugar and rice until an avalanche occurs. First, smooth the surfaces of the materials so that they are both horizontal. Then slowly tilt the bowls until a layer of grains shears off and runs down the slope. You'll find that there is a critical angle, called the angle of maximum stability, at which sliding takes place. And when the avalanche has finished, the slope of the grains in the bowl will have decreased to a stable value (Fig. 4.14). This is called the *angle of repose*. However high a pile of grains grows, recurrent landslides ensure that the slope stays more or less constant, equal to the angle of repose. These 'avalanche angles' depend on the grain shape: they are bigger for rice than for sugar, whereas granulated sugar, caster sugar and couscous (all with roughly spherical grains) have similar angles of stability within the accuracy of this kitchen-table demonstration.

It was quite by chance that Hernán Makse decided to conduct his initial experiments with sand and sugar, which have slightly different grain *shapes* and therefore slightly different angles of maximum stability and repose. Different grain *sizes* alone would not have produced stratification, but only crude segregation with the larger grains gathered at the foot of the slope. So in the model that he and his colleagues developed, they tried crudely to mimic this difference in shape. They considered two types of grain, square and rectangular. These drop on to the pile and stack in columns (Fig. 4.15). This is a good example of what 'making a model' means in physics, because of course the grains in the experiments are clearly not squares and rect-angles, nor do they stack up in regular vertical columns. The skill resides in deciding whether such simplifications matter. Here, the researchers judged that they would not wreck the model's ability to mimic what is going on.

The heap was assigned characteristic angles of repose and maximum stability—θ_r and θ_m. When a grain drops on to the pile to create a local slope greater than θ_m, it tumbles down from column to column until it finds a position in which the slope is less than or equal to θ_m. If the slope is everywhere equal to θ_m then the grain tumbles all the way to the bottom. This is a sign that the slope is primed for a landslide. Makse and colleagues stipulated that if a grain rolls all the way to the foot of the pile, then *all* the grains at the slope surface tumble, starting at the bottom, until the slope everywhere is reduced to the angle of repose θ_r.

This model recreates the striped landslides (Fig. 4.15e) Because the large grains are 'taller' and so tend to make the slope steeper than the small grains, they tumble more readily. This accounts for the segregation of grains, with the larger ones ending up at the bottom. Stratification, meanwhile, arises from the different grain shapes and thus the different angles of repose and max-imum stability. This simple model doesn't capture everything that is going on in the experiments: for example, to get good stratification you also need to get the pouring rate right (not too fast). That has something to do with the details of how grains collide, which is not given much attention in the model.

ROLL OUT THE BARREL

As bricklayers know, you can mix up powders inside the rotating drum of a cement mixer. But that is generally done with water added. If the powders are dry, perfect mixing might never happen no matter how many times the drum turns. This became clear to Julio Ottino and his

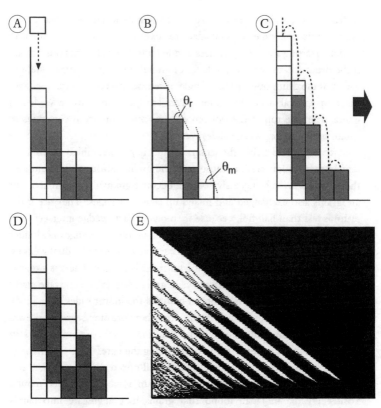

FIG. 4.15 A simple model for how stratified landslides occur. This assumes that the two types of grain have different shapes: square and rectangular. As they are poured, they simply stack into columns, with all the rectangular grains standing upright (a). The difference in height between one column and the next is not permitted to exceed three times the width of the square grains: this defines the angle of maximum stability θ_m. And when a landslide occurs, it relaxes the slope so that the height difference between adjacent columns nowhere exceeds twice this width. This defines the angle of repose θ_r (b). If a new grain added to the top of the pile creates a slope greater than θ_m, it tumbles from column to column until it finds a stable position. If this takes it all the way to the bottom of the pile (c), then the entire slope undergoes a landslide until the slope is everywhere equal to (or less than) θ_r (d). Although highly stylized, the model generates the same kind of segregation and stratification seen in the experiments (e). (Image e: Hernán Makse, Schlumberger-Doll Research, Ridgefield, Connecticut.)

colleagues at Northwestern University in Illinois when they tried in this way to mix two types of salt, identical except for being dyed different colours, starting with the powders divided into two segments (Fig. 4.16a). If the drum rotates slowly, the layer of granular material remains stationary until the drum tips it past its angle of repose, whereupon the top layer slides in an avalanche (Fig. 4.16b). This abruptly transports a wedge of grains from the top to the bottom of the slope. The drum continues to rotate until another wedge slides.

With each avalanche, the grains within it get scrambled (there's no stripey segregation here, because only the grains' colour is different). So the powders in each successive wedge become gradually intermixed. But are grains also exchanged and mixed up *between* wedges? They are if the drum is less than half-full, because then bits of each wedge intersect with others (Fig. 4.16b). But when it is exactly half-full the wedges no longer overlap (Fig. 4.16c), and mixing then occurs only *within* individual wedges. If the drum is more than half-full, the outcome is striking. There is a region around the outer part of the drum where avalanches and mixing take place, but in the central region is a core of material that never slides (Fig. 4.16d). The initially segregated grains in this core therefore stay segregated even after the drum has rotated many times (Fig. 4.16e). In theory you could spin this cement mixer for ever without disturbing the core.

Even if you start with a barrel full of well-mixed grains, they will not necessarily stay that way when tumbled. In 1939 a researcher named Yositisi Oyama in Japan found that grains may segregate into bands when turned in a rotating cylindrical tube (Fig. 4.17a).* This happens if the grains have different angles of repose—for example, tiny glass beads will separate from sand. And this banding will happen for different-sized grains even if they have the same angle of repose, if they are rotated in a tube with a succession of bulges (Fig. 4.17b). The larger balls gather in the necks if the tube is more than half full, but in the bellies if it is less than half full. In all these segregation processes it is important that the grains only

---

*The phenomenon did not become general knowledge until much later, for Oyama's paper seems to have left little impression until it was rediscovered in modern work on granular materials. Julio Ottino suggests that it was an example of what the biologist Gunther Stent called 'scientific prematurity': a discovery made too early to be connected to facts and theories then known.

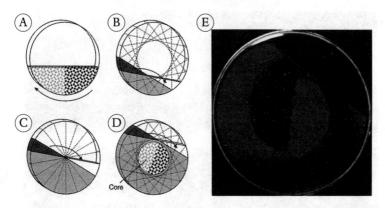

FIG. 4.16 Avalanches of grains in a rotating drum will mix grains that are initially divided into two segments (*a*). As the drum turns, there is a succession of avalanches each time the slope exceeds the angle of maximum stability (*b*, *c*, *d*), transposing the wedge shown in dark here to the wedge shown in white. If the drum is less than half full (*b*), the wedges overlap, and so the two types of grain eventually become fully mixed. If the drum is exactly half full (*c*), the wedges do not overlap, so mixing takes place only within individual wedges. When it is more than half full (*d*), there is a central core in which avalanches never take place, so this circular region never gets mixed. This unmixed core is clearly visible in experiments (*e*). (Photo: Julio Ottino, Northwestern University, Evanston, Illinois.)

partially fill the tube, so that there is a free surface across which grains can roll in avalanches.

Joel Stavans from the Weizmann Institute of Science in Israel and his coworkers have proposed that this 'Oyama effect' (as it deserves to be, but has not yet become, known) might be used to separate different kinds of grains in a mixture. They say that the banding results from the complex interplay between two properties of the grains: their different angles of repose and the differences in their frictional interactions with the edge of the tube. When the researchers created a model of the tumbling process based on these assumptions, they found it predicted that the well-mixed state of the grains is inherently unstable, since small, chance imbalances in the relative amounts of the two grains are self-amplifying. A tiny excess of one grain type in one region grows until that part of the column contains only that type exclusively.

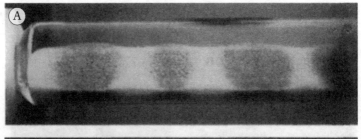

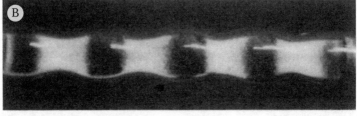

FIG. 4.17 Grains of different shapes (and thus angles of repose) will segregate into bands when rotated inside a cylindrical tube (a). Here the dark bands are sand, and the light bands contain glass balls. In a tube with an undulating cross-section (b), a difference in size alone is enough to separate the grains, which segregate into the necks and bellies. (Photos: Joel Stavans, Weizmann Institute of Science, Rehovot.)

That cannot be the whole story, however, since the segregated patches seem to have a fairly well-defined typical size even in a tube of uniform width (as opposed to one where the size is set by width oscillations of the tube). Amplification of random imbalances, in contrast, would give patches of all sizes. How, then, to account for this characteristic length scale in the pattern? Stavans and colleagues point out that this is analogous to what happens when a mixture of miscible ('mixable') fluids is suddenly made immiscible (for example, by cooling them down). This happens in rapid 'quenching' of a mixture of molten metals below their freezing point: the two metals separate into blobs of more or less uniform size. In this process, called spinodal decomposition, blobs of all sizes grow in a self-amplifying manner, but those of a certain size are more stable than the others and therefore get selected preferentially. By carefully controlling the conditions, such as the rate of cooling, a particular blob size can be obtained by design; this is often done in metallurgy and chemical engineering to make particles of specified sizes.

SELF-ORGANIZED AVALANCHES

The problem with landslides and avalanches is that you can never quite be sure when they will happen. Sometimes they are set off by unpredictable disturbances such as earthquakes—that is how most tsunamis are generated, when seismic tremors send subsea sediments sliding down a slope. But avalanches seem also to have an intrinsic capriciousness. For simple piles of more or less identical grains, we can feel sure that there is trouble in store once the slope exceeds the angle of repose; but even then it is hard to know how big the landslide will be. If the grains are of many different sizes and shapes, or if they have complex frictional properties (as is the case for wet soil or sticky snow crystals), or if the surface on which they rest is rugged, we cannot be sure what to expect. All we know is that when grains are set in motion, we had better watch out.

Does that mean we can make no useful predictions about the timing or the size of avalanches? Not exactly. Rather, it means that avalanche science, like earthquake science, is necessarily statistical: we can't say exactly what *will* happen in a particular event, but we can know about the relative probabilities of what *might* happen.

And in fact, studying landslides this way has proved to be astonishingly productive. For it seems that the humble pile of sand is analogous to a great many 'catastrophic' processes that happen in nature, from forest fires to ecological mass extinctions. The key feature of all these processes is that, while they are unpredictable, they are not fully random in the sense that each event happens independently of the others. There is a subtle but very important statistical regularity in such processes, and it is one that connects these seemingly disorderly, unpredictable phenomena to ones that give rise to well-defined patterns. For it appears that landslides, like most of the patterns we have encountered already, are *self-organized*.

To see what that means, let's go back to the simple conical sand pile. In 1987, physicists Per Bak, Chao Tang, and Kurt Wiesenfeld at Brookhaven National Laboratory on Long Island, New York, devised a model to describe the way such a pile behaves as it grows by new grains being poured on to the apex. These researchers initially had no intention of studying piles of sand. Rather, they were investigating an aspect of the electronic behaviour of exotic solids that is so recondite I do not even propose to describe it. What they came to realize is that the behaviour of the electrons in these materials can be represented by the behaviour of sand grains in a pile. That is not to say that the electrons themselves form a pile, or anything of the kind. It is a little like the way one can model oscillating chemical reactions by thinking of foxes eating hares, as I described in Book I: the equations describing both behaviours look the same.

Bak, Tang, and Wiesenfeld then considered a pile of sand grains with a specific angle of repose onto which new grains are steadily dropped. In the simplest version of this model the sand pile is two-dimensional—as if, in the experiments on striped landslides, that I described earlier, the two glass plates are so close together that the pile is only one grain thick. What happens as grains are added to this pile one by one at random points?

The pile builds up unevenly, so that its slope varies from place to place (Fig. 4.18a). But if anywhere this slope exceeds the angle of maximum stability, a landslide is triggered which sweeps down the hillside and reduces the slope everywhere to less than this critical value (Fig. 4.18b). How big is the landslide? That is to say, how many grains does it set tumbling? Bak and colleagues found that, in their simple model of a sand pile, a single new grain added to the pile can induce a landslide of any magnitude. It might set off just a handful of grains, or it might bring about a catastrophic sloughing of the entire pile. And there is no way of telling, no matter how carefully we inspect the slope beforehand, which it will be.

In other words, the smallest perturbation can have an effect quite out of proportion to its size—or it can remain just that, a small perturbation with a small effect. There is no characteristic *scale* to the system: in this case, no typical or favoured number of grains is set tumbling when one more is added. The model sand pile is thus said to be *scale-invariant*. We will see later in this book and in Book III that this is a common characteristic of a certain class of disorderly forms and patterns: they have no natural length scale, so that we cannot be sure whether we're looking at the whole system or just a small fragment of it.

Yet while landslides of all sizes (from a single grain to the whole slope) are possible, they are not all equally probable. If we keep track of the

FIG. 4.18 The slope of a granular heap varies locally from place to place (*a*). In this pile of mustard seeds, small variations in slope are superimposed on a constant average gradient. When the slope approaches the angle of maximum stability, the addition of a single seed can trigger an avalanche (*b*). This avalanche can involve any number of grains, from just a few to the entire surface layers of the slope. (Photos: Sidney Nagel, University of Chicago.)

number of landslides of different sizes as we continue to add grains to the slope, we will find that there are many more little slides than big ones. Landslides in which the whole slope tumbles are rare indeed. Thus, the number of landslides decreases as the number of grains it involves increases. In the model of Bak and colleagues, this relationship has a particular mathematical form: the frequency f (or equivalently, the probability) of an avalanche is proportional to the inverse of its size s (Fig. 4.19). This kind of inverse relationship between the size of an event and the probability that it will attain that size is commonly called a $1/f$ ('one over f') law. Technically, it is an example of a so-called *power law*. That simply implies that some quantity y is proportional to some other quantity x raised to some power a: $y \propto_2 x^a$. Here, the power (also called the exponent) a is equal to -1: $f \propto_2 s^{-1}$. I hope this maths is not too distressing—it is basically the same as that which I introduced at the start of Book I, where I promised that it was all you would need here. We will return to power laws again in Book III. The key point is that a $1/f$ law is *not* what one would predict if each of the events it is describing were each happening independently of the next. In that case, the mathematical relationship between size and frequency or probability is instead described by the familiar bell curve.

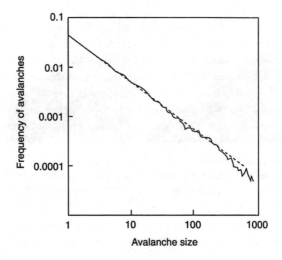

FIG. 4.19 In a simple model of sand-pile avalanches, the frequency of an avalanche of a certain size decreases in inverse proportion to its size. On a graph plotting the logarithm of frequency against the logarithm of size, this relationship appears as a straight line with a slope of minus 1 (shown by the dashed line). (After Bak, 1997.)

Avalanches can be regarded as *fluctuations* of the sand pile: disturbances of a steady state, corresponding here to a slope inclined at the angle of repose. If we added no more grains to such a slope, it would remain unchanging indefinitely. When we do add grains, the slope has a continual series of convulsions (avalanches) of all sizes, all of which rearrange the slope so that on average it rests at the angle of repose. The constant rain of grains makes the sand pile a *non-equilibrium system*. We encountered these in Book I, where I explained that they are the sources of most natural patterns. We also saw there that, in order to keep a system out of equilibrium, we have to provide a constant supply of energy, and generally of matter too. That is what is happening to the sand pile: the falling grains are perpetually injecting energy and matter into the slope. This is what drives the fluctuations.

It turns out that $1/f$ laws govern the size of fluctuations observed in a wide variety of natural and human-made systems. An electrical current flowing through a resistor undergoes tiny fluctuations of this kind, and so does the amount of heat and light (the luminosity) emitted by the Sun. These latter fluctuations are due to outbursts of super-hot plasma called solar flares, which are generated by writhing magnetic fields in the Sun's outer atmosphere. The light from distant astrophysical objects called quasars shows the same kind of variability; so do some records of volcanic eruptions and of rainfall. Some palaeontologists have argued that the geological record of mass extinctions (catastrophic events that wipe out a significant fraction of organisms on Earth) seem also to show a $1/f$ relationship between the size and frequency of the events, at least for marine ecosystems. In all these cases, we can see abrupt avalanche-like events happening on all size scales.

Although $1/f$ power laws have long been known to govern the statistical behaviour of fluctuations in diverse natural systems, the reasons for it were unknown before Per Bak and his colleagues devised their model of a sand pile. Here, then, was a simple instance of such behaviour, in which all the ingredients were known, that might offer some clues about the general origin of $1/f$ laws.

There is something very peculiar about this model sand pile: it is constantly seeking the least stable state. We are used to quite the opposite: nature generally seems to crave stability, which is why water runs downhill, golf balls drop into holes, trees topple. The sand pile, however, is for ever returning to the state in which it is poised on the brink of an avalanche. Each time an avalanche occurs, this precarious balancing act gives way; but then as further grains are added, the system creeps right back to the brink.

States like this, which are susceptible to fluctuations on all scales at the slightest provocation, have been known to physicists for a long time. They are called *critical states*, and are found in systems as diverse as magnets, liquids, and theoretical models of the Big Bang. Every liquid adopts a critical state at a particular temperature and pressure, called the critical point. If you heat a liquid, it evaporates to a vapour once it reaches boiling point: the state of the fluid changes abruptly from a (dense) liquid to a (rarefied) gas. But above the critical temperature this abrupt change of state no longer happens; instead, the fluid passes smoothly and continuously from a dense liquid-like state to a diffuse gas-like state as its pressure is lowered. The critical point is the point at which there is no longer any sharp distinction between 'liquid' and 'gas', and no boiling point separating the two.

At the critical point of a fluid, its density fluctuates wildly from one place to another (Fig. 4.20). In some patches the fluid has a liquid-like

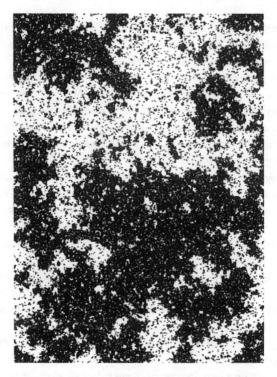

FIG. 4.20 At the critical point of a liquid and a gas, the distinction between these two states of matter disappears. A critical fluid contains variations in density on all scales, with domains of liquid-like fluid coexisting with domains of gas-like fluid. Here I show a snapshot of such a fluid, obtained from a computer model of a critical point. The dark regions represent liquid-like, dense domains, and the white regions are gas-like. (Image: Alastair Bruce, University of Edinburgh.)

density, in others it is gas-like, and these patches are constantly changing without any characteristic size or shape: they are ephemeral fluctuations. The fluid is poised right on the brink of separating out into distinct liquid and gas regions (which is what it does at temperatures below the critical point), but it cannot quite make up its mind to do so. Now, if we were to decide whether each little pocket of fluid at the critical point will be 'liquid-like' or 'gas-like' by tossing a coin, it wouldn't look like an actual critical state but would seem more uniformly random, lacking any big patches of one kind or the other. The reason the density fluctuations look the way they do is that each pocket of fluid is affected by the state of the fluid around it—these patches are not all independent of one another.

It is extremely difficult to hold a fluid in the disorderly yet interdependent mixture of liquid-like and gas-like patches of a critical state. It is on the brink of separating into two big regions, one liquid-like and the other gas-like. A critical point is thus like a needle balanced on its tip: theoretically a perfectly balanced state does exist, but it is unstable against even the slightest nudge or breath of air. But while the theoretical sand piles of Bak and colleagues have this same critical character, being susceptible to fluctuations (avalanches) on all length scales provoked by the smallest perturbation (the addition of a single grain, say), they seem in contrast to be robust. That is to say, instead of constantly seeking to *escape* the critical state, the sand pile seeks constantly to *return* to it—like a needle that constantly wobbles but never falls. The researchers called this phenomenon *self-organized criticality*, reflecting the fact that the critical state seems to organize itself into this most precarious of configurations.

Bak began to see signs of self-organized criticality just about everywhere he looked. In a theoretical model of forest fires, for instance, fires can grow to any size, burning just a few trees in the immediate vicinity or spreading catastrophically over large areas. The fires thus spare clusters of unburned trees of all sizes as they sweep through the forest. If the trees re-grow slowly, the forest is maintained in a self-organized critical state by occasional fires. And it has been known for over four decades that earthquakes follow a $1/f$ power law (or something very close to it), called the Gutenberg–Richter law: earthquakes occur on all scales of magnitude, from a shelf rattler to a city leveller, with the probability declining as the magnitude gets larger. This smacks of self-organized criticality, and a simple mechanical model of how geological faults slip past one another reproduces this power-law behaviour.

Per Bak believed that in self-organized criticality he had uncovered 'a comprehensive framework to describe the ubiquity of complexity in Nature'—and not just in nature but in human systems too, such as the fluctuations of economic markets and the spread of technological innovation. There is no doubt that many of the *models* developed to describe complex systems of this sort do find their way into a self-organized critical state. But verifying that the real world also behaves that way is much harder, and many of the claimed instances of self-organized criticality, such as in mass extinctions and forest fires, remain contentious. One of the problems is that the statistics are often ambiguous. To be sure you are seeing a particular kind of mathematical relationship between size and frequency, and not just something that looks a bit like it over a small range of size scales, you need a lot of data—which you can't always get. There may not have been enough mass extinctions since the beginning of the world, for example, to allow us to be sure that evolution operates in a self-organized critical state. Another problem is that, whereas in a model you can usually be sure exactly what all the important parameters are, and can see the effect of changing each one independently, in reality complex systems may be susceptible to all manner of perturbing influences, some more obvious than others. Will a model of earthquake faulting that includes a more realistic description of the sliding process or of the geological structure of the Earth still show self-organized criticality?

In fact, it is even contentious whether real sand piles, the inspiration for the original model, have self-organized critical states. You might imagine that this, at least, ought to be a simple experiment to perform: you just drop sand grain by grain on to a pile and observe how big an avalanche follows from each addition. But there is no unique way to measure the size of an avalanche, and the experiments that have been conducted so far do not give a clear answer. For example, Sidney Nagel and his colleagues at the University of Chicago found in 1989 that real sand piles seem always to undergo large avalanches, in which most of the top layer of sand slides away, while other experiments in the early 1990s seemed to generate power-law behaviour like that expected of self-organized criticality. Many researchers now think that real sand piles don't show true self-organized criticality; but it is also possible that it is subtly disguised so that it doesn't show up in the measured size of avalanches.

If sand piles are not really self-organized critical states, perhaps that is not so surprising, since real sand is not like model sand. For one thing, the grains are not identical in size, shape or surface features, and these

microscopic details determine how readily they slide over one another.*
And the collisions of grains dissipate energy in a way that is not accounted
for in the simplest models. In 1995 Jens Feder, Kim Christensen, and their
co-workers at the University of Oslo in Norway attempted to settle the
debate about whether or not grain avalanches are examples of self-organized criticality. They added a new twist to the tale: instead of studying sand
piles, they looked at piles of rice. This was because rice grains do not roll or
slip over one another as readily as sand grains do (just as rugby balls do not
roll as well as footballs), and so they capture more accurately the behaviour
of grains assumed in the computer models that do clearly show self-
organized criticality (a rare example of an experiment being adapted to
fit the model rather than vice versa). The grains tumble if they exceed the
angle of repose, but moving grains quickly stop rolling when this is no
longer so. The researchers looked at two-dimensional piles in which the
rice grains were confined to a narrow layer between two parallel glass
plates (Fig. 4.21).

Observing enough avalanches to provide trustworthy statistics was a
slow and tedious process, and took about a year. But at the end of it all, the
researchers concluded that the behaviour of these granular piles depended
on the kind of rice that they used: specifically, on whether it was long
grain or short grain. Long-grain rice with a larger ratio of length to width
seems to show true self-organized critical behaviour, with a power-law
relationship between the size of the avalanche (the researchers actually
measured how much energy each one released) and its frequency of
occurrence. But short-grain rice, which is more nearly spherical and so
more like sand, showed different behaviour: instead of a simple power
law, the relationship between size and frequency was more complicated
than a $1/f$ law. That relationship could, however, be easily mistaken for a

---

*Per Bak's book on self-organized criticality, boldly titled *How Nature Works*, contains a
photograph of an experimental sand pile built up against glass walls that unwittingly refutes
the point it is meant to illustrate: that grain avalanches are a 'scale-free' process with no
characteristic size scale. The slope shows clear chevron stripes of darker grains, presumably
generated by the stratification process I discussed earlier (pages 92–95). This is a reminder,
perhaps, that scientists risk overlooking things when they don't expect to find them, even
when they are obvious in the data.

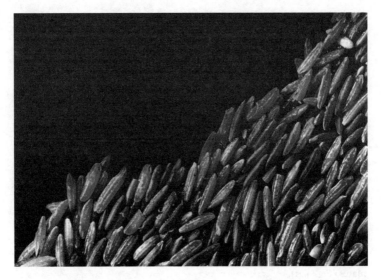

FIG. 4.21 A section of a rice pile confined between two glass plates. Notice how uneven the slope is at this fine scale. (Photo: Kim Christensen, University of Oslo.)

power law (and thus for the signature of self-organized criticality) if the measurements were not taken over a wide enough range of avalanche sizes, possibly explaining why others had previously claimed to have seen self-organized criticality in experimental sand piles.

So, while piles of grains apparently *can* show self-organized behaviour, they will not necessarily do so, and indeed generally will not; it depends on (amongst other things) the shape of the grains and how their energy is dissipated during tumbling. This both vindicates and modifies Bak's assertions: self-organized criticality seems to be a real phenomenon, not just a product of computer models, but it may not be universal or even particularly easy to observe or achieve. For the present time, sand piles appear to be an intriguing but limited metaphor for nature's complexity.

THE RISE OR FALL OF NUTS

I am not very good at using up the last of a packet of muesli. It is dreadfully wasteful, I know, but the fact is that by the time you get to the bottom, all the large pieces of fruit and nuts are gone and all that's left is a rather unappetizing residue of dry oat flakes. The big pieces always seem to stay on top, and the small ones settle to the bottom. This has become known to physicists as the Brazil nut effect.

The sorting of grains of different sizes in a shaken granular medium is well known to engineers, but the reason for it is still disputed. You might think that shaking would simply mix up grains of different sizes, but clearly this is not so—usually, the larger grains instead rise mysteriously to the top. Even if the packet of muesli left the factory well mixed (which is unlikely), the Brazil nuts and banana flakes are likely to have reached the top by the time the packets have made their way by rumbling juggernaut to the supermarket. The British engineer John Williams of the University of Bradford studied this effect systematically in the 1960s. He saw a single large particle rise up through a bed of finer powder as it was vibrated up

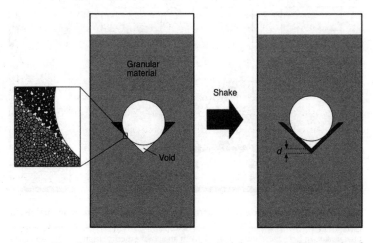

FIG. 4.22   How the brazil nut rises? Here there is a single large grain within a powder of smaller grains (I have exaggerated the differences in size, for clarity). The large grain (white) tends to accumulate an empty space below it. When the box is shaken vertically, the large grain jumps away from the walls of the void, allowing the smaller grains in a ring around it (of which we can see two wedge-shaped cross-sections here, in dark grey) to slide down into the void. So when the large grain settles again, it comes to rest on the cone of dark grains and so has risen by a small distance *d* equal roughly to the thickness of the dark layer. In this way, the large grain gets ratcheted steadily upwards with each jump.

and down. Williams suggested that the large particle is ratcheted upwards: as all the particles jump up during a shake, the large one leaves a void beneath it, into which smaller grains fall (Fig. 4.22), so the small grains prevent the large one from settling back to its original height after each shake. In 1992 the physicists Remi Jullien and Paul Meakin observed this ratchet process in computer simulations of the shaking process.

But there is more than this to the inexorable rise of the Brazil nut. Sidney Nagel and his colleagues have conducted experiments in which they shook glass beads, all the same size except for one or a few larger ones, in a glass cylinder. Again, the big beads gradually rose to the top. Nagel's team tracked the motion of individual beads by dying a layer in which many small beads surrounded a large one near the bottom of the cylinder. The large bead rose up vertically, accompanied by the small beads immediately around it; but the dyed beads at the edges of the layer,

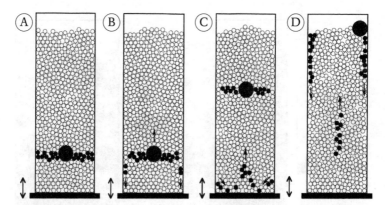

FIG. 4.23 Grains in a tall column undergo convection-like circulating motion: those in the centre rise upwards, and those at the edges crawl downwards in a narrow layer. The images shown here were reconstructed from an actual experiment in which some glass beads were dyed so that their movements could be tracked. An initially flat layer of dyed beads near the bottom of the column (a) separates into downward-moving beads at the edges and rising beads at the centre (b). At the top, the latter move outwards to the walls, and then begin to descend (d). The beads at the edges, meanwhile, move inwards to the centre and start to rise, once they have reached the bottom (c,d). A single large bead gets trapped at the top because it is too large to fit in the narrow downward-moving layer around the column's edges. This convective motion thus causes the segregation of different sizes. (Images: Sidney Nagel, University of Chicago.)

in contact with the container's sides, began instead to make their way *down* to the bottom of the container (Fig. 4.23). As the central group of beads continued to rise, those that descended at the edges reached the bottom and then began to rise up again in the centre. And once the large bead and its surrounding small companions reached the top, the large bead stayed there but the smaller ones made their way to the sides of the container and began to travel downwards.

Thus, the small, dyed beads are in fact circulating: rising at the centre and descending at the edges, just like fluid in the convection cells we encountered in the previous chapter. The size segregation here is merely a by-product of this convection-like motion: larger beads are pushed up on the rising column of the cell but, once at the top, are unable to follow the cycle further because the descending portion of the cell is confined to a very thin layer (about the thickness of the small beads) at the container's edge.

So apparently, grainy materials don't just flow but *convect*. In fact this has been known for a long time; Michael Faraday seems to have seen it in 1831. But what drives the flow? In normal fluids, we saw earlier that convection is a result of buoyancy due to density differences between layers of the fluid at different temperatures. But all the particles in Nagel's granular medium have the same density—they are all (with the exception of the large bead, which doesn't have to be there for convective flow to occur) the same size. Nagel and colleagues worked out that the important factor is the frictional force between the beads and the walls of the container, which hinders the upward jumps of the peripheral beads during each shake. In support of this idea, they found that more slippery walls reduced the circulatory motions, while rougher walls made them more pronounced. (As the computer simulations of Jullien and Meakin involved no walls at all, they would not have seen these convective effects.)

'Brazil nut' particles do not always rise to the top of a mix of shaken grains—sometimes the big grains sink to the bottom. This 'reverse Brazil-nut effect' was predicted in 2002 by Daniel Hong at Lehigh University in Pennsylvania and his colleagues, and was confirmed the following year in shaken beads of metal, wood, and glass by researchers at the University of Bayreuth. Hong's theory suggested that, in mixtures of two types of spherical bead, the big beads should switch from going up to going down at particular thresholds of the size and density ratio of the small and large beads. The German team confirmed that this theory usually gives the right prediction, but that the behaviour also depends on how fast and how hard the mixture of beads is shaken. Segregation may also happen in thin layers of grains shaken horizontally, like sand swirled in a gold-panner's sieve: depending on the relative densities of the beads, the larger ones can become marshalled either around the edges or in the centre of the layer.

All of which suggests that the only way to really tell what your cereal packet will do is to shake it and see.

JUMPING BEANS

When Michael Faraday first shook the packet, he saw both circulatory (convective) motion of grains and the spontaneous appearance of heaps or 'bunkers' on the surface of the material. He suspected that the air that is present in the tiny spaces between the grains plays a role in these effects. When a layer of grains is shaken vigorously, the bottom of the layer jumps away from the floor, creating a cavity that contains a lower no air pressure. The abrupt difference in air pressure between the gas amongst the grains and this almost air-free cavity pushes some grains underneath the pile as it rises, and so creates unevenness in the layer, leading to heaping. By performed experiments in which the pressure of the gas permeating the granular layer is changed systematically to investigate the effect on heaping, researchers have recently confirmed Faraday's mechanism.

Faraday proposed it as an explanation for the patterns seen in vibrating grains by the eighteenth-century German physicist Ernst Chladni. Chladni found in 1787 that if fine sand is scattered over a metal plate and the edge of the plate is bowed with a violin bow to excite an acoustic vibration, the powder gathers into lines and spots that interweave in rather beautiful patterns (Fig. 4.24). The vibrations of the plate depend on how they are excited: how hard, and at what frequency. Like a kind of two-dimensional organ pipe or guitar string, the plate has particular 'modes' of vibration in which whole numbers of waves fit perfectly onto the surface. As with a guitar string, some points of the vibrating surface will be moving upwards while other parts move downwards, and at some point in between there is a so-called node where the surface doesn't move at all (Fig. 4.25). On a

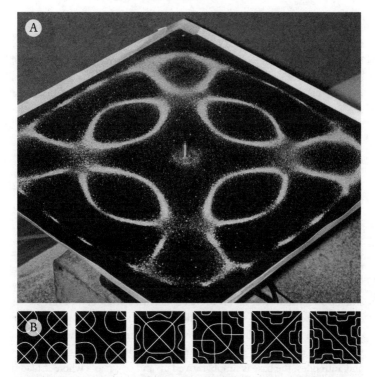

FIG. 4.24   Chladni figures form in fine powder scattered on top of a metal plate, when it is vibrated with a violin bow (a). The range of different figures is immense—a small selection is shown in b. (Photo a: Biological Physics Department, University of Mons, Belgium.)

plucked string this node is a single point, but on a plate it can trace out a line. Chladni noted that fine particles and coarse particles sitting on the surface behave differently: the fine ones pile up at the 'antinodes', where the plate has the maximum amplitude of up-down motion, while the coarse ones gather at nodes where there is no plate motion. Faraday suggested that, while the big grains just jump about until they reach the points on the surface that aren't moving, the fine particles are pushed to the antinodes by air currents caused by the pressure differences induced as the layer jumps and opens up cavities. He tested this by sticking pieces of paper to the plates to block or channel the air flow, and found that these

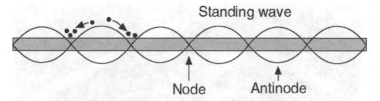

Standing wave

Node   Antinode

FIG. 4.25 The patterns are caused by the movement of grains towards nodes, where the vibrations of the plate create no displacement either up or down. Some grains may instead move towards antinodes, the regions of maximum displacement.

had the expected effect of redirecting the grains. Surprisingly, Faraday's general idea did not receive clear experimental support until 1998.

His theory suggests that the behaviour of a layer of vibrated grains should depend on the air pressure above it. In the mid-1900s, Harry Swinney and colleagues Paul Umbanhowar and Francisco Melo at the University of Texas at Austin decided to see what happens to a shaken layer of grains when there is no air at all, and thus no air currents. They studied a very thin layer of tiny bronze spheres, each about the same size as a typical sand grain, in a shallow, sealed container which was pumped free of air and vibrated rapidly up and down. The vibration here was uniform: rather than exciting the surface acoustically so that it is laced with nodes and antinodes, the base is simply moved up and down as a whole. Chladni figures trace out the shape of the acoustic waves, which serve as a patterning template. But here there is no such template: no forces that would obviously drive the grains into specific patterns.

And yet there was pattern in abundance. In fact, this set-up has proved to be the most fertile breeding ground for grainy patterns so far known. The granular layer became organized into a series of dynamic ripples: stationary waves in which the grains are constantly rising and falling in step with each other. These wave patterns can be visualized by 'freezing' the little bronze balls at one point in their motion using a stroboscope: the light flashes on and illuminates the balls in step with their oscillatory rise and fall, and so always catches the pattern at the same point in its cycle. Swinney and colleagues saw patterns that might by now be familiar to us: stripes (interspersed with dislocations), spirals, hexagonal and square cells, and more random, non-stationary cell-like patterns that appear to be turbulent (Fig. 4.26).

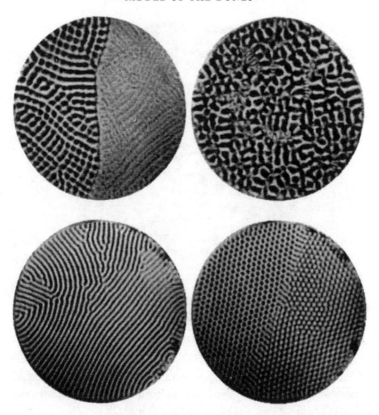

FIG. 4.26 When shaken vertically, a shallow layer of grains can develop complex wave patterns, including stripes, squares, and hexagons, as well as less orderly, 'turbulent' patterns. (Photos: Harry Swinney, University of Texas at Austin.)

The pattern depends on the frequency and amplitude of the shaking, and switches between patterns happen abruptly as critical thresholds are crossed. The grains move up and down at frequencies that are simple ratios of the shaking frequency: once in every two shakes or, for larger amplitudes of shaking, once in every four. But different parts of the same pattern may oscillate out of step with one another, so that one part is rising while the other is falling. Then the strobe light catches the grains out of phase, illuminating peaks (bright) in one region and troughs (dark)

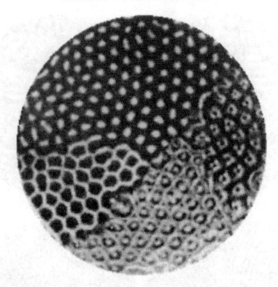

FIG. 4.27 When different domains of the granular pattern rise and fall out of step, the stroboscope that captures a frozen image reveals the domains at different stages in their cycle, and so the patterns look different, even though they are in fact identical. (Photo: Harry Swinney, University of Texas at Austin.)

in another (Fig. 4.27). When the amplitude of shaking exceeds a certain threshold, the patterns dissolve into disorder: if the grains are thrown too high, they can no longer organize all their motions in step.

Why doesn't the layer of grains simply move up and down as a whole, without any pattern at all? Well, it does do just that if the amplitude of vibration is small. But above a critical amplitude there is a *bifurcation* in this flat layer: a switch from a single steady state of grain motion to two states, one in which the grains are rising and one in which they are falling. Each of these states may exist at different points in the layer, which becomes organized into alternating stripes that rise and fall out of step. At low frequencies the stripes criss-cross in a square pattern. Then, at a second critical amplitude, a second bifurcation occurs, leading to a hexagonal pattern. At one point on the oscillatory cycle this pattern appears as an array of little spot-like peaks, whereas if the stroboscope is set up to capture the pattern half a cycle later one sees an array of hexagonal honeycomb cells, with voids in the centre. Thus the pattern reveals itself as a doubled-up oscillation: spot, void, spot, void... You can make out both of these patterns (along with two intermediate configurations in out-of-step regions) in Fig. 4.27.

These patterns result from collisions between grains: this is what puts the grains literally 'in touch' with one another, so that their movements can become synchronized. Swinney and colleagues found that they could reproduce the patterns in a model if they assumed that the grains lose a little energy when they collide. At the same time, of course, they are being pushed up by the base of the container, and pulled back down by gravity. But remarkably, we don't even need to take that into account to explain the patterns: it all just depends on collisions, which is to say, on the horizontal movements of the grains. Troy Shinbrot, then working at Northwestern University in Illinois, has shown that this is so. In his model the grains do not move vertically, but are simply given a little kick in a randomly chosen horizontal direction on each vibration cycle, reflecting the randomizing influence of shaking. This may bring a grain into collision with one of its neighbours, whereupon they lose a little of their energy. There seems to be nothing in this prescription but a recipe for randomness, and yet after just a hundred shakes Shinbrot found an initially random scattering of grains organizing themselves into stripes, hexagons, and squares (Fig. 4.28). Which pattern is selected depends on the strength of the randomizing effect of shaking and on the average distance that each grain travels before colliding with another. As well as reproducing the patterns observed experimentally, Shinbrot found others that had not been seen before (Fig. 4.28d) but which might, he suggested, appear if the right experimental conditions could be found.

All these patterns can be regarded as wave-like disturbances of a uniform system: the pattern emerges much as a standing wave appears in a vibrated tray of water.* But there is another way in which we can describe them: as ranks of individual elements that interact with each

---

*This is more metaphor than analogy. A standing wave has its wavelength set by the size of the container: the wave has to 'fit'. But the regular patterns in vibrated grains are more striking because the wavelength is determined by the properties of the grains themselves: by how they collide, and how far they move between collisions. This is why these latter are truly *self-organized* patterns.

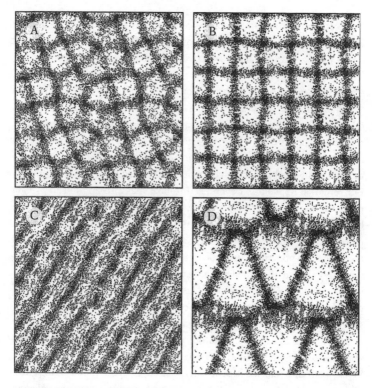

FIG. 4.28 Complex ordered and irregular patterns arise spontaneously in a simple model of a shaken thin layer of grains as a result of the interplay between random 'noisiness' of the grain movements and collisions between them. (Images: Troy Shinbrot, Northwestern University, Evanston, Illinois.)

other to form ordered structures, like people arranging themselves evenly spaced out on a crowded beach. You could regard this picture as a kind of 'particle-like' alternative to the wave-like picture of global instabilities with characteristic wavelengths. It is possible to capture and study these 'pattern particles' of shaken sand in isolation. Swinney and colleagues found that, for a certain range of layer depths and shaking frequencies and amplitudes, they could generate just a few lone oscillating peaks, or even just a single one, in the granular layer (Fig. 4.29).

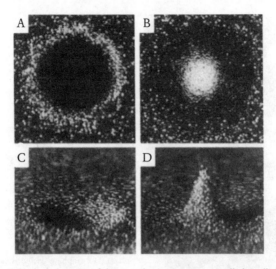

FIG. 4.29 Individual elements of the granular wave patterns, called oscillons, can be isolated. Each oscillon is a single peak that rises and falls, as seen here from above (*a,b*) and from the side (*c,d*) at different points in the cycle. (Photos: Harry Swinney, University of Texas at Austin.)

They call these lone peaks oscillons: isolated 'packets' of oscillation. An oscillon is a peak of jumping balls at one instant, and a crater-like depression the next. It looks rather like the splash made when something plops into a puddle of water—except that the splash doesn't die out in a series of spreading ripples but keeps jumping back up as if captured in a time loop. These beasts exist for a shaking amplitude slightly smaller than that required for the appearance of the full pattern. Swinney and colleagues discovered that they could conduct a curious kind of 'oscillon chemistry' with them. Oscillons can move around through the granular layers, and when they encounter one another, one of two things can happen. Each oscillon jumps up and down at half the shaking frequency, and so two oscillons must be either in step with each other or perfectly out of step, one rising to a peak when the other makes a crater. Two out-of-step oscillons attract each other, like particles of opposite electrical charge, enabling them to link up into 'molecules' (Fig. 4.30*a*). Whole strings of

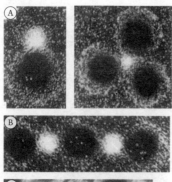

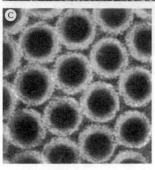

FIG. 4.30 Oscillons behave like particles that attract one another if their oscillations are out of phase, whereas they repel one another if they are in phase. Out-of-phase oscillons can form 'oscillon molecules' (a) or chains (b). A group of in-phase oscillons will pack together in an orderly, hexagonal arrangement with an equal distance between them (c). (Photos: Harry Swinney.)

out-of-step oscillons can form, analogous to chain-like polymer molecules (Fig. 4.30b). The range of the attractive interaction is only small, about one and a half times the width of an oscillon, so they have to approach quite closely before sticking together. Oscillons that are in step, meanwhile, repel each other, like particles with the same electrical charge. A party of in-step oscillons will form a hexagonal pattern (Fig. 4.30c), since this allows each oscillon to stay as far from all of its neighbours as possible. This is like a fragment of the global hexagonal pattern in Fig. 4.26.

THE WORLD IN A GRAIN OF SAND

Shaking, tumbling or even simple pouring can therefore cause grains to mix, to unmix, or to form rather wonderful patterns. I don't imagine for a moment that we know all there is to know about the capacity of granular substances for generating spontaneous patterns, nor have I been able here to survey all of those that are currently known. Because at present there is no general 'theory of grains' that will tell us what to expect, and because not even the scientists studying them have yet acquired the necessary intuition, we cannot predict what we might see in a given experiment.

Osborne Reynolds, the doyen of fluid dynamics whom we encountered in Chapter 2, had big visions for grainy flows. Reynolds discovered that, in order for a collection of grains to flow, it must expand a little. Left to settle, the grains will pack together densely, and then there is nowhere for them to go. This seems all very reasonable, but Reynolds deduced from it a rather extraordinary conclusion. He decided that somehow this 'dilatancy' of powders could explain all the mechanical behaviours in nature. At the turn of the century, when there was very little understanding of the internal structure of atoms, no one could be sure what space and matter looked like at the subatomic scale. Reynolds decided that it was in fact filled with grains: rigid particles that, by his estimate, were about five million-trillionths (5×10^{-18}) of a centimetre across—much smaller than a proton. The notion of all these submicroscopic grains rubbing up against one another conjures up an image strikingly similar to that envisaged by René Descartes in the seventeenth century, who postulated a universe filled with vortices in contact. Descartes's fluid, you might say, became Reynolds's powder.

His idea was rather eccentric even by the standards of Victorian science. John Collier's portrait of Reynolds from 1904 shows him holding a basin of ball-bearings, and in a prestigious lecture two years earlier he revealed what he had in mind with this seemingly innocuous collection of hard particles: 'I have in my hand the first experimental model universe, a soft India rubber bag filled with small shot.' Blake's phrase is invoked to the point of cliché in research on grainy materials, but for Reynolds it became a reality.

5

FOLLOW YOUR NEIGHBOUR

Flocks, Swarms, and Crowds

I once saw the Harvard biologist E. O. Wilson asked a challenging question by a nervous young man at a public lecture. Wilson won a Pulitzer prize in 1990 for a book written with fellow naturalist Bert Hölldobler on the behaviour of ants—a definitive and weighty tome, yet communicated with joy and passion. But honestly, the young man exclaimed with awkward and genuine puzzlement, how could a person spend their entire research career studying something as straightforward, as prosaic and, let's face it, as *small* as ants?

Wilson is by no means famous only for his work on ants. As a leading proponent of sociobiology, otherwise known as evolutionary psychology, he became notorious in the 1970s for his endorsement of what some considered (wrongly) to be a rigidly deterministic view of human nature with right-wing overtones. Nonetheless, his enthusiasm for ants runs very deep, and the audience was perceptibly anxious at how Wilson might receive this apparently naive dismissal. But there was no hint of condescension or defensiveness in his answer. He simply suggested to the young man that he scatter some sugar near an ant's nest on a warm day—and then sit back and watch. What will unfold, he implied, will soon seem an appropriate subject for a life's work.

You could say much the same about many other organisms. Sit in a park at dusk and watch a group of swallows dive and swoop, and you enter into a profound mystery. Watch schools of fish evading their predators off the Great Barrier Reef, or herds of wildebeest crossing the savannah, or even observe a culture of bacteria proliferate and spread under a microscope, and you begin to understand that biological organization does not stop at the level of the individual organism. All these groups display motions that hint at some grand scheme, some sense of coherence and even purpose that governs the collective behaviour of the community.

Biologists have long appreciated the significance of group behaviour in the animal kingdom, among which they recognize the coordination exhibited by creatures such as ants and bees—so-called *social* insects—as something special and remarkable. But only rather recently have these collective motions been seen as something akin to *flow*. In the previous chapter we saw how flow may take place among solid particles (grains) and how this can lead to remarkable forms and patterns. Now I want to look at what happens when those grains become animated: when they are moved not simply by gravity, or wind, or shaking, but under their own steam, propelled by wings, legs, fins, or wiggling bodies. By making our 'grains' self-propelled, we might imagine that their motions will degenerate into the random, disorder of a crowd. Yet flocks and swarms show that this need not be so: even animal crowds (including those composed of humans) may display coherence and order.

LAWS OF MOTION

This coherence may sometimes be so pronounced that it seems miraculous. How does each bird in a flock sense what all the others are going to do, so that they all change direction simultaneously (Fig. 5.1)? Perhaps they don't—maybe there is a single leader that all the others follow? Careful observation of flocking shows that the motions are not executed in perfect synchrony (an idea that has forced some researchers in the past to suppose a form of thought-transference or mental communication via electromagnetic fields). Rather, changes in direction seem to propagate rapidly through the flock like waves. The same is true of fish: in the early 1970s, Russian ecologist D. V. Radakov identified waves of 'excitation' that pass like ripples through schools.

This suggests that the manoeuvres are being triggered by individuals and copied by those around them. Yet the waves seem to pass from one

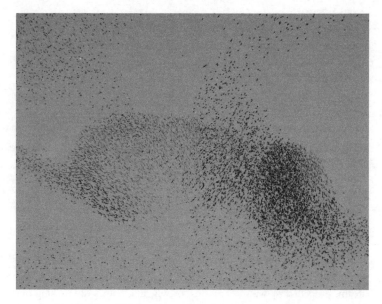

FIG. 5.1 The flocking of birds seems to be a near-miraculous feat of coordination. (Photo: Jef Poskanzer.)

bird to the next faster than their speed of reaction permits. In 1984, Wayne Potts of Utah State University proposed that birds watched the wave approaching and then timed their manoeuvre to coincide with the wave's arrival. He compared this to the timing of high kicks by a chorus line, although the phenomenon is probably more familiar today from the Mexican waves conducted by football crowds. The problem is that this places a weighty burden on the sensory capabilities of birds: they need to be able to spot the wave from afar and then to judge precisely when they need to move in order to synchronize their motion with the wavefront as it reaches them. And who are the leaders? Studies of group motions have generally sought in vain to identify specific individuals who tell the others what to do. Besides, if each manoeuvre involves a different 'leader', how does the group decide which of them it will be?

We now know that the collective motions of animal flocks and swarms do not demand leaders at all. They appear to be self-organized: the coherent group behaviour emerges from simple, purely local interactions between individuals, who have no sense of what the whole group is doing and no ability to perform great feats of anticipation. This understanding arose initially not through any fundamental desire to unlock the secrets of group motion, but out of an attempt merely to imitate such motion in a computer model. The software engineer Craig Reynolds at the Symbolics computer company in California was accustomed to thinking about problems of computer representation and animation in terms of algorithms and rules rather than recondite theories of physics or biology: if you want to generate a certain kind of behaviour on the computer, what rules must you follow? In 1986 he set about mimicking the coordinated motion of bird flocks and fish schools in systems of 'particles' that move through a simulated landscape. After watching blackbirds flock, he decided that each bird was merely responding to what its neighbours did. What rules would reproduce this behaviour on the computer?

Reynolds attributed to his 'particles', which he called boids (a condensation of 'bird-like droids'), three basic behaviours that governed their 'steering':

1. Avoid collisions or close encounters with flockmates.

2. Align with the average direction of your neighbours.

3. Stick together: steer towards the average 'centre of gravity' of your neighbours.

Which individuals are the boids' 'neighbours'? Reynolds assumed that this included all boids within a specified radius, which typically extends for just a few 'boid-widths'. Each boid thus heeds only those nearby.

These rules seem guaranteed to ensure that the boids form groups: the third rule acts rather like an attractive force which binds them together. They also seem designed to make the boids line up and execute roughly parallel motions. What they do not do is include any explicit prescription for the kind of large-scale coordination familiar in real flocks: one might imagine that they merely prescribe bunching into little clusters. But when Reynolds ran simulations on the computer according to these rules, the motion of his boids looked uncannily like the real thing (Fig. 5.2).

Reynolds was not too bothered about whether his rules were biologically realistic: he just wanted the simulations to look right, because the

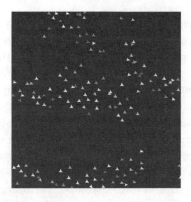

FIG. 5.2 A snapshot of a flock of 'boids' moving according to local rules based on those devised by Craig Reynolds. This simulation uses the NetLogo software devised by Uriel Wilensky and colleagues at Northwestern University. It can be downloaded free of charge at <http://ccl.northwestern.edu/netlogo> and contains many sample programmes for studying flocking and other pattern-forming behaviours in animal populations and biological ecosystems.

ultimate goal of his computer programme was to supply a tool for computer animation. To that end, he happily added in other rules that made the results look even more realistic, regardless of whether there was any biological justification for them. And indeed they looked wonderful, which is why they have been used in several films, such as the bat swarms in *Batman Returns*. But scientists became alerted to the implications when Chris Langton, a researcher on complexity at the Santa Fe Institute in New Mexico, found out about Reynolds's work. He invited Reynolds to speak at a 1987 workshop on artificial life, where boids were recognized as a classic example of 'emergent behaviour': self-organization produced from local rules that dictate the interactions of individual agents.

Research on 'artificial life'—computer simulations that generate life-like behaviour—is sometimes criticized for itself being little more than a sophisticated form of computer gaming, preoccupied with reproducing appearances without any concern about the fundamental reasons for them or whether they reflect what might be happening in the real world. All the same, Reynolds's boid model held an important message: collective motion does not require global vision, nor does it need complex behavioural origins. It is enough simply to follow your neighbour. This perspective was taken up by physicists and biologists in the 1990s in attempts to formulate more rigorous theories of flocks and swarms. In 1994, Tamás Vicsek and his student András Czirók at Eötvös University in Budapest teamed up with researchers at Tel Aviv in Israel to devise a theory of collective motion. While they were aware that their model might have something to say about birds and fish, its primary motive was to explain collective motion in a much simpler kind of organism: bacteria, specifically colonies of *Bacillus subtilis*. The Tel Aviv team, led by Eshel Ben-Jacob, had found that *B. subtilis* can grow into complex patterns, some of which we will encounter in Book III. Among these are branching tendrils, some with curling tips that resemble a kind of plant (Fig. 5.3*a*). Ben-Jacob and his colleagues inspected these under the microscope and found that the curls are produced when the bacterial cells line up and move in arcing filaments. Other branching patterns were tipped with blobs of cells (Fig. 5.3*b*) that, on close inspection, turned out to be rotating vortices (Fig. 5.3*c*).

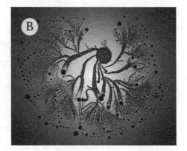

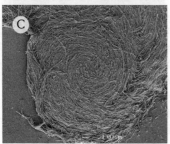

FIG. 5.3 Some of the complex branching patterns formed by bacteria (*a*, *b*). The curling tendrils in *a* are formed by the alignment of cells into streams, while the blobs at the ends of branches in *b* are circulating vortices of cells, visible in the electron microscope (*c*). (Photos: *a,b*: Eshel Ben-Jacob and Kineret Ben Knaan, Tel Aviv University; *c*, Colin Ingham.)

What causes these coordinated, circulating streams of cells? We saw in Book I that some micro-organisms, such as the bacteria *E. coli* and the slime mold *Dictyostelium discoideum*, communicate with one another by releasing and sensing chemical compounds that diffuse into the environment. They move towards larger concentrations of this chemical signal, a type of behaviour called chemotaxis. Single-celled organisms are by no means alone in responding to chemical signals of this sort: higher animals do it too, guided by hormones called pheromones. The impulse does not have to be chemical: some organisms will move towards sources of heat or light, for example. The basic principle is that an organism moves in the direction that improves its circumstances: towards warmth, nutrients, or its fellows.

Some organisms seem to achieve this sort of directed motion by 'deciding' whether or not to execute a turn depending on whether or not this makes the conditions better or worse. This kind of motion is called klinotaxis, and is performed by some species of fish. A group can often sense and follow changes in an environmental signal if individuals don't simply look for that signal themselves but also respond to what one another is doing—turning if their neighbours turn, for instance. This can help to prevent individuals from wandering off course, and can communicate the 'scent' to individuals who haven't discovered it for themselves. Collaborative klinotaxis seems to help fish schools to migrate to warmer or cooler waters over huge distances that involve only very gradual changes in ocean temperature.

Exactly *how* individuals in a group communicate with one another is a complex matter. They might send and receive chemical signals, say, or see directly what others are doing, or they may merely get turned around and lined up in one another's slipstream. Vicsek and his colleagues did not worry about the mechanisms; they simply assumed that these kinds of interaction happen, and that, as with Reynolds's boids, they obey a set of simple rules. In fact, there was in this case only a single rule: each 'self-propelled particle' (SPP), travelling with a constant speed, moves in the direction of the average motion of neighbours lying within some fixed distance. There was just one other ingredient in this SPP model: the motion of each particle also had a random element, a sort of tendency to get disorientated. If this randomness, or 'noise', is too strong, it can reduce the SPPs to a collection of randomly jiggling particles, each ignoring the others, like the molecules of a gas (Fig. 5.4a). If the noise is lessened, however, then the particles start to become aligned, and when they are not too densely packed into a given space, they form little flocks that execute collective movements in random directions, with some tendency to form circulating gyrations (Fig. 5.4b). If the particles are more densely crowded, and the noise is low, these groups cohere into a collective motion of the whole group, travelling in a single direction (Fig. 5.4c).

Is this how flocks form? One of the predictions of the SPP model is that coherent, self-organized motion emerges spontaneously, and suddenly, once the density of the animal group exceeds a particular threshold. That is a physicist's neat version of events: you tweak a dial and look at the

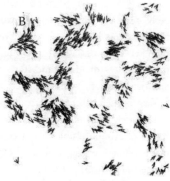

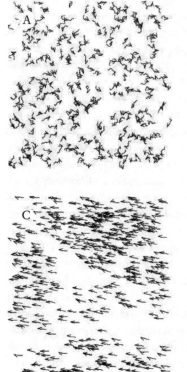

FIG. 5.4 Self-propelled particles interacting via local rules that produce alignment with neighbours may show collective behaviour. If the particle motions contain too strong an element of randomness ('noise'), there is no coherence (a). But if the noise is reduced, the particles gather into aligned bunches (b), or, if their density is large enough, all stream together in a single direction (c). (Images: Tamás Vicsek, Eötvös Loránd University, Budapest.)

effect. It isn't easy to find such things realized in nature. But Jerome Buhl of the University of Sydney and his collaborators have arranged it in the laboratory. They looked at the group behaviour of desert locusts (*Schistocerca gregaria*), a scourge of agriculture in Africa, the Middle East, and Asia. Desert locusts have repeatedly devastated crops in these vulnerable regions for centuries, bringing famine and misery in the wake of plagues of biblical proportions. Once a swarm forms, it is almost impossible to control. Swarms may contain tens of billions of locusts, covering more than 1,000 square kilometres, and may travel up to 200 km a day if the wind is favourable. Before such a swarm of mature adults takes to the air, it is heralded by the formation of 'marching bands' of wingless juvenile insects: great columns several kilometres long. These bands arise from the aggregation of smaller groups, which gather together and march collectively into new territories, gathering recruits. If a group fails to swell its numbers this way, however, it will eventually disband. The vital issue, then, is how and when a group ceases to behave as a collection of meandering individuals and starts to display coordinated motion, signalling the onset of a swarm. Understanding this process could hold the key to disrupting a nascent swarm before it can cohere.

In marching bands there are typically around 50 locusts on every square metre of ground. Buhl and his colleagues wondered whether the switch from solitary to gregarious, coherent behaviour in desert locusts might happen at a threshold density like that seen in the SPP model. So they studied how the behaviour of the insects changed in the laboratory as their density increased. They placed juvenile locusts in a ring-shaped arena and watched how their 'marching' changed as more and more insects were added to the group, altering the density from around 12 to around 295 per square metre. At low densities, the creatures wandered at random. But once the density reached between 25 and 60 locusts per square metre, the insects began to align themselves, tramping around the ring-shaped arena in an orderly fashion. The direction of this motion changed suddenly every hour or so. Above 74 locusts per square metre, however, these direction changes ceased, at least over the eight-hour observing period: there was a single, relentlessly circulating flow of locusts, like a true marching band. It looked, then, as though the SPP model predicted just what the researchers found.

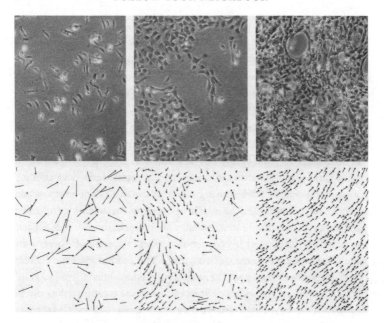

FIG. 5.5 Keratocyte cells show a switch from random (left) to coordinated (right) movement as their density increases. The lower images show individual cell motions deduced from video data, of which the top images are snapshots. (Images: Tamás Vicsek, Eötvös Loránd University, Budapest.)

Vicsek's group has also seen such a switch from disordered to ordered motion in colonies of cells called keratocytes, found in human skin and other tissues (the Hungarian group studied those of goldfish scales). These cells are able to move across a surface—that is how they gather together in the right places to form tissues. Vicsek and colleagues found that as the density of cells increased, they saw a change from random motion to collective, vortex-like motions and finally to coherent flow of the whole group, just as seen in the SPP model (Fig. 5.5).*

*It's not clear that mobile single cells will always show these sharp switches from random to collective motion as they become more densely populated, however. A team of US researchers found that *Bacillus subtilis* bacteria seem to show only gradual changes in motion—for example,

GROUP MEMORY

The SPP model has its limitations. For one thing, it doesn't rule out collisions between the 'particles', whereas real creatures seek to avoid bumping into one another. Nor does the model have any ingredient for forming groups that stay permanently together: it can generate little swarms that form and reform, and it can make all the particles in the box move as one, but it does not explain how a group such as a fish school can hold together without any box to contain it.

Iain Couzin of Princeton University and his co-workers have tried to correct these deficits, while looking at motions in three-dimensional space rather than just two. The local rules governing the behaviour of their particles are more complicated, but they are biologically reasonable. Each individual is surrounded by concentric, spherical zones of interaction, which determine the creature's behaviour when another individual enters one of these zones (Fig. 5.6a). Very close to the individual there is a zone of repulsion: when a neighbour enters this zone, the creature will perform an avoidance manoeuvre to ensure they don't collide. Beyond this there is a zone of orientation: when other individuals come within this zone, the creature will aim to align its motion with the average of theirs. And beyond that is a zone of attraction: the creature will simply try to stay close to others within this zone, without bothering about orientation. It prioritizes these rules, so that, for example, it will forget about orientation and concentrate on collision-avoidance if there are any individuals in the zone of repulsion.

Exactly how a group of simulated creatures behaves in any particular case depends on all the various ingredients of the model, such as how dense the group is, how big the zones of interaction are, how fast the creatures move, and so forth. But even though these offer countless permutations of the model 'settings', there are just four different basic types of behaviour that emerge (Fig. 5.6b–e). One is a group that holds together but without otherwise any coherence of motion. This is like a

in the average velocity of cells—as the density of a colony increased. They suspected that here the kind of sudden changes predicted by the SPP model were smeared out by random 'noise' in the system, for example because of the swirling fluid medium or the range of different cell sizes.

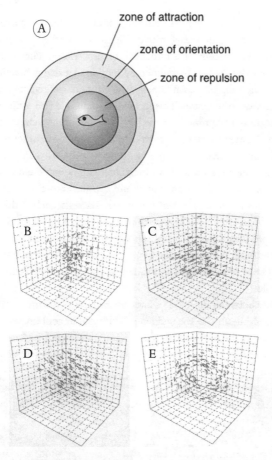

FIG. 5.6 In the model of collective animal motion devised by Iain Couzin and his co-workers, each individual moves in a manner that takes into account the presence of others in three nested spherical zones (a). The innermost zone is repulsive: the creature will aim to avoid others that come this close. But it will orientate its motion with the average of others inside the second zone. And it will merely try to stay close to others within the outermost zone of attraction. Four different types of group behaviour emerge from these rules (b–e). The group may cluster without any orientational alignment, in a swarm (b). Or the motions may be aligned, like a flock or school (c). Or it may all move in a single direction (d), or circulate in a hoop (e). (After Couzin and Krause, 2002.)

swarm of insects such as gnats, buzzing about at random in a dense cloud. Then there is a group that moves hither and thither in approximate alignment, reminiscent of a bird flock or fish school. The group might alternatively display highly regimented motion in a single direction, like a group of migrating birds. And finally, the group may circulate together in a doughnut or torus shape. This last might seem odd and artificial, but in fact it is rather common: some fish, for example behave this way (Fig. 5.7a). Certain species of fish have to keep moving in order to breathe; the torus allows them to do that without actually travelling anywhere. Moving in collective rather than individual circles may help to conserve energy, since adjacent fish may then move inside each other's slipstream, reducing the frictional drag of the water. Bacteria exhibit this toroidal motion, too: this is what is happening in the vortices of *Bacillus subtilis*, for example (Fig. 5.3c), and it has also been seen in colonies of the slime mould *Dictyostelium discoideum* (Fig. 5.7b), where it is thought that the cohesion between cells (the equivalent of the zone of attraction) plays a crucial role.

Switches between these states happen abruptly as the model conditions are changed, rather like the sudden switch from uncoordinated to coherent motion in the SPP model. This is a common property of collective

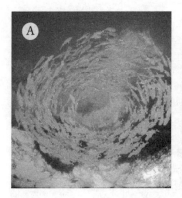
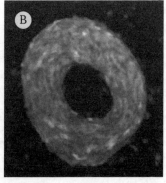

FIG. 5.7 Toroidal group motion is surprisingly common. It has been seen, for example, in fish (a), and slime moulds (b). (Photos: a, Copyright Norbert Wu; b, Herbert Levine, University of California at San Diego.)

modes of behaviour: it is what we find, for instance, when a liquid freezes. Cooling water just a fraction of a degree below freezing point transforms it all to ice: the change in conditions is tiny, but the consequent change in the state of the molecules is marked and global. Freezing and melting are the result of interactions between all the constituent molecules in the material, and they are known to physicists as examples of so-called *phase transitions*. The models of collective motion in animals also seem to show phase transitions.

Couzin and colleagues found that switches between the stable states— say, from swarm-like to flock-like motion—did not always happen at the same point for both directions of change. So two groups that started in a swarm and a flock state might remain in those respective states even as the conditions they experience—their group density, say—change until they are identical. In physics this kind of persistence of a collective state is known as hysteresis. Couzin and colleagues call it 'collective memory': the group behaviour depends not only on the conditions it experiences and the rules it observes, but also on its own history. We will encounter other examples of patterns that display this historical contingency.

An array of collective states of motion in animal groups might have adaptive benefits, helping the organisms to cope with a changing environment. For one thing, such motion allows information to be transmitted rapidly throughout the group. If a few individuals detect a predator and take evasive action, this sets up a ripple effect that travels quickly from neighbour to neighbour until those far away 'feel' the effect of the threat even though they cannot see it for themselves. In fact, both fish and birds do seem to become better aligned when a predator is near: by shifting to this highly coordinated motion, they create conditions under which disturbances can propagate like waves through the crowd. When Couzin and colleagues added the rule that individuals perform an avoidance manoeuvre of any predator that comes into their range of detection, while stipulating that the predators move towards the highest-density part of the group, they were able to reproduce many features of the escape response of real fish schools, such as the sudden expansion of the group around a predator, possibly creating an empty space around it, and splitting of the group into smaller groups (Fig. 5.8).

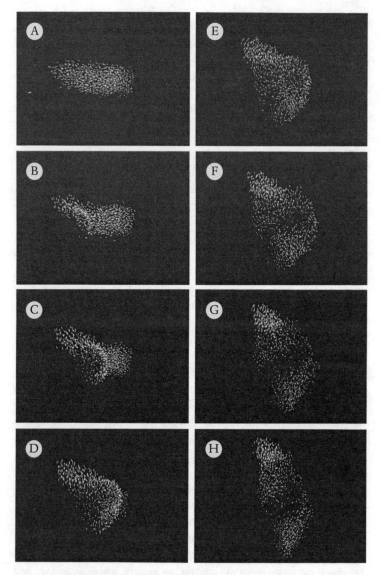

FIG. 5.8 Snapshots of a computer simulation of the escape manoeuvre of a fish school evading a predator. (Images: from Couzin and Krause, 2002.)

These apparent similarities between the model and nature are encouraging. But one of the big unanswered questions is whether the behavioural rules of the model really do correspond to those used by the animals. This is extremely hard to test, because it is not easy or obvious to work backwards from the observed collective behaviour to the rules that generate it. One study has suggested that fish respond to the motion of only a small number of neighbours (about three) when carrying out an escape move. But it's not at all clear that the fish are 'counting' here, rather than simply reacting to all of their neighbours within a particular distance.

A crowd of fish is no more likely to be uniform than a crowd of people. They may differ in age, size, speed, and manoeuvrability, and also in their intrinsic preferences and tendencies. How does this diversity affect the group behaviour? Do some individuals prefer to take up specific positions in the flow, for example? Being at the front of a group has its attractions: if the group finds food, you get the first bite. But there are also obvious drawbacks: if a predator finds you, then you are apt to *become* the first bite. So edge positions might be favoured by the boldest individuals, or simply by the hungriest, who may be content to risk greater vulnerability for the chance of greater reward. A position in the group centre, meanwhile, is not necessarily more secure: you're sheltered from predators, but you're also more hindered from escaping if a predator comes steaming through the crowd.

Models of collective motion have shown that differences in behaviour can cause individuals to adopt particular positions even without any 'conscious' choice to do so. If the rules that govern an individual's motion differ slightly, this can automatically move them to different positions, often causing segregation into subgroups that share the same rules. For example, differences in swimming speed owing to different body sizes can sort a group of fish into large and small. A naive observer, seeing this, might think that big fish actively seek out other big fish and so on—and in fact this is often what biologists have assumed when they see segregation within populations. But in fact the sorting may emerge without that kind of complex decision-making at all. All the same, the outcome may have evolutionary benefits. Parasites that infect fish might spread themselves more readily by altering their hosts' behaviour, for example making fish slower or less manoeuvrable so that they adopt positions more likely to get them eaten. Conversely, a group may ruthlessly shed individuals that stand out from the rest when predators are nearby, making the group less likely to draw unwelcome attention—not because the fish are truly heartless or prejudiced against difference, but because the local behavioural rules they follow differ for 'normal' and 'odd' individuals in a way that makes the segregation happen.

FOLLOW THE LEADER

These self-organizing motions seem to deny any need for leaders. But sometimes a few individuals really do know best—for example, if they have discovered where food is to be found. When the group's motion becomes collective, with each member responding to its neighbours, it becomes much easier for privileged information of this sort to be shared among the group for its collective benefit. Couzin and his co-workers have investigated how this allows moving groups to make complex decisions in an efficient manner. When they adapted the model described above so that a certain proportion of individuals all moved in the same preferred direction, only a small fraction of such 'informed' members were needed to guide the whole group in that direction. And as the group became bigger, this proportion got smaller: the 'accuracy' of the group improved. Real swarms do seem to have this ability: a swarm of honeybees can find a new nest site with only one in 20 individuals knowing the way to a good location. The key point here is that these agents who 'know the right way' do not actually have any way of letting the others know they have this information—no one in the group knows who 'knows best', or even if there are any informed individuals at all. But the slight bias on the group motion imposed by a small number of individuals who all head in the 'good' direction is enough to carry the others along.

What, however, if the group contains a mixture of informed individuals with different, conflicting information? Couzin and colleagues found that in such cases a consensus always emerges, even if the different leadership groups are of equal size. With two groups of leaders, each of which has a different preferred direction, the consensus direction depends on how much these choices differ. The group selects the average of the two directions, unless these are too diametrically opposed (more than about 120° apart), in which case the direction of the largest leadership group is chosen. In this last case, if the two groups are of exactly the same size then one or other of the directions is selected at random.

An averaging consensus might not seem ideal: by selecting the average of the two 'leadership' directions, the group will be heading towards neither of the leaders' goals. But that is not necessarily a problem. If the two targets are a long way off, averaging takes the group in the right general direction towards them both, and as it approaches, the angle of divergence between the two possible routes gets steadily bigger (just as you can hold two distant objects in your vision at once, but may have to turn your head from one to the other once you get closer). When this divergence exceeds the critical angle of 120°, then the group chooses to head for one target or the other.

In any event, the main conclusion is that the model shows how a group of interacting individuals can respond to the information gathered by just a few, and can reach a collective decision about how to use that information even without any sophisticated means of assessing and discussing it. In this sense, animal groups seem to have a democratic capacity that we might reasonably deem somewhat enviable.

CROWD PSYCHOLOGY

It is one thing to make models of how fish and birds move, for it seems likely that they are governed primarily by instinct. Using a computer model of boids or self-propelled particles constrained to follow a few simple rules in a robotic manner seems to do not too great an injustice to these creatures' undoubted complexity. But might people move in this collective, herd-like fashion, too?

That seems a bold, even an impertinent, thing to suggest. But it is surely a whole lot less impertinent than the approaches taken in some of the earliest work on the motions of human groups, wherein so little intelligence was attributed to individuals that they were reduced to so many dumb, inanimate particles: the notion was that human crowds could be regarded as genuine fluids. The physicist James Lighthill, an expert in fluid dynamics, proposed in the 1950s that road traffic was rather like water flowing down a pipe, and together with Gerald Whitham at Manchester University he attempted to use hydrodynamics to develop an understanding of the vagaries of motion on a busy highway and how it is affected by bottlenecks and junctions. And in the 1970s an Australian engineer named L. F. Henderson wrote about 'crowd fluids', considering them as though they were collections of randomly moving particles, like those of a gas, and charting the statistical distributions of different walking speeds.

Figuring out how people move through space does not seem to have been felt as a strong priority by social scientists, who have found it rather more appealing to study how groups acquire customs and traditions, behavioural traits and ideas and fashions. But the problem of movement is of immense importance to architects and town planners, who need to know where to lay out walkways for maximum convenience, where to place emergency exits and how many to use, how to avoid crushes in dense crowds, and many other mundane but very real problems linked to the way we use space.

Motivated by the sociological notion of 'social forces' that govern behaviour, Dirk Helbing and Péter Molnár of the University of Stuttgart developed a model of pedestrian motion in the mid-1990s that posited the operation of forces of attraction and repulsion between individuals. They didn't mean to imply that these interactions exist in the same sense as they do between magnets or electrically charged plates. What they meant was that we tend to act in a crowd as though such forces exist. In particular, we avoid collisions, as if a repulsive force holds us back from bumping into each other. Two people walking towards each other will veer aside as they approach (and if you think we are smarter than particles, bear in mind that particles never make the same choice as one another and so end up colliding anyway).

What else controls our movements around open spaces? In general, we are trying to go somewhere: we have a particular destination in mind, and will move towards it along something like the shortest path. (Really the shortest? No, not necessarily. In a dense crowd we cannot know what the shortest collision-avoiding path will be; and in other situations our steps might be diverted by other factors, as we will see.) We each have our own preferred walking speed, and will speed up until we attain this, unless something prevents us from doing so.

These were the simple ingredients of Helbing and Molnár's model. You can see that it assumes rather uncomplicated, single-minded pedestrians.

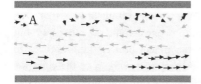

FIG. 5.9 Computer simulations of pedestrians moving in opposite directions down a corridor show that they organize themselves into counterflowing streams, even though there is no explicit prescription for this in the 'rules' of the model (*a*). Such behaviour is commonly seen in real life (*b*). (Photo *b*: Michael Schreckenberg, University of Duisburg.)

On a crowded day in the shopping precinct, that is perhaps not a bad description of what we are. What sort of crowd motion, then, do these rules produce? Helbing has applied the model to a wide range of situations: negotiating busy intersections, for example, or passing through unmarked doorways. One of the simplest situations places the walkers in a corridor, travelling in both directions. (The corridor could as easily be a pavement.) If the crowd density is high, there is apparently a recipe here for chaos and congestion. What emerges, however, is a surprising degree of order (Fig. 5.9*a*).

The pedestrians arrange themselves into counter-flowing streams, trailing in each other's footsteps. This might not seem surprising at all, for obviously it makes sense to follow the person in front of you: that way, you are far less likely to collide with someone coming the other way. But there is no rule to that effect in the model prescription—nothing to generate 'following' behaviour. This simply appears once the rules are enacted. Of course, it is possible, indeed likely, that we do show an active tendency to follow one another and form streams—if such an impulse is included in the pedestrian model, then the bands form more quickly, and in fact are evident even before two counter-flowing groups meet. But the point is that this ingredient is not essential for streaming to occur: it emerges from the propensity for collision-avoidance. There is ample evidence that streaming really does occur, and you have no doubt seen it for yourself (Fig. 5.9b).

ANT MOTORWAYS

This sort of 'lane formation' among walkers is not unique to humans. It is strikingly exhibited by army ants (*Eciton burchelli*), voracious carnivores that carry out truly terrifying raids on their arthropod prey. Bands of several hundred thousand individuals may form trails many metres wide and over 100 m long leading from their colony to the prey. The raid has to be finished by dusk, since the ants are inactive at night, and so they have no time to lose. The trails are divided into lanes that are used selectively by individuals setting out from the colony or returning back carrying their prey (Fig. 5.10).

Ants certainly may follow where others have gone before. This is one of the key aspects of their foraging behaviour: each ant marks its path by laying down a pheromone, and other ants are drawn towards this chemical trail. They seek out the highest pheromone concentration, which tends to keep them travelling along a previously marked path rather than wandering off on their own. Thus, trails become self-reinforcing: the more ants that travel that way, the more strongly the route is chemically marked, and so the more likely others are to use it. This reinforcement ensures that prey are located efficiently: once a few successful individuals have found the way there, their pheromone footsteps will guide others. If, on the other hand, a small group of ants gets separated from the nest and so loses its

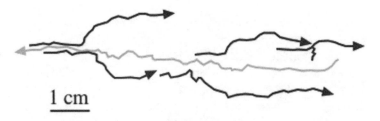

FIG. 5.10 Three-lane traffic in army ants, *Eciton burchelli*. The paths of five outgoing ants are shown in black, and those of a single returning ant in grey. The former use two 'outside' lanes, while the latter stays to the centre. (After Couzin and Franks, 2002.)

return path, it may mill around aimlessly in a circle, each following the other and not realizing that the path leads nowhere (Fig. 5.11).

This trail-following behaviour gives rise to characteristic branching patterns in army-ant raids (Fig. 5.12a), which can be mimicked by computer models that include the self-amplifying mechanism of pheromone release (Fig. 5.12b). How, though, do the trails get divided into lanes designated for advancing and returning workers (Fig. 5.10)? Iain Couzin and his colleague Nigel Franks at Bristol University think that the answer lies in the ants' wish to avoid collisions, particularly with those coming in the other direction.

Army ants do not have good vision (they are in fact almost blind), but they can literally feel the presence of others that come into physical contact. They also have antennae that extend their tactile field in front of them. Couzin and Franks assumed that if other ants come within this field, an ant will change direction, veering off its initial course. When not manoeuvring out of the paths of others, the ants seek out and follow a pheromone trail, moving to where the concentration is highest.

There are two other rules governing their motion. First, the ants know whether they are setting out from the nest or returning to it. Real army ants do seem to know this too, although it is not clear how—if they are made to perform a U-turn along their route, they will quickly turn again to resume their original direction. Second, the researchers made outgoing and incoming ants different in just one crucial way: the former had a greater propensity to change course when they came into contact with

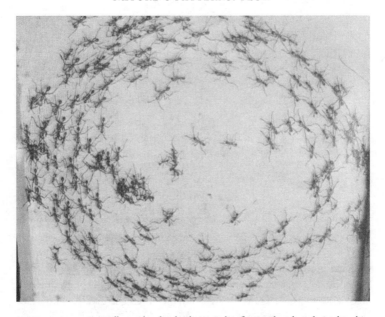

FIG. 5.11 Army ants will march relentlessly in circles if a circular obstacle is placed in their midst. They are simply following each other's pheromone trail, unaware that it leads nowhere. (Photo: from T. C. Schneirla, *Army Ants* [W.H. Freeman, 1971], kindly supplied by Nigel Franks, University of Bristol.)

others. It is not clear whether real army ants behave this way, but it makes sense: ants laden with prey are less manoeuvrable and so less likely to change direction.

When the model was run with a single, straight pheromone trail laid down to guide the ants, they became segregated into three lanes, with the inbound ants at the centre and the outgoing ones on either side. This is precisely the arrangement found in nature (Fig. 5.10); it is also seen in foraging termites. Again, there is nothing in the rules to 'tell' an ant whether to use the inner or outer lanes—this emerges spontaneously from their interactions with one another. And once more, lane formation makes good sense, because it means that the paths of all the ants are less likely to be disrupted by collisions. But the ants don't need some instinct that tells them 'when leaving the nest, keep to the outside lanes'. That will

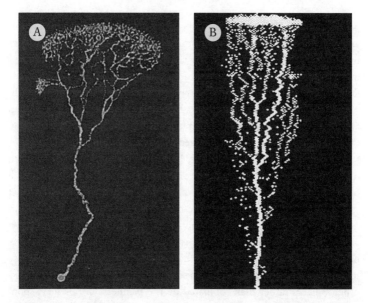

FIG. 5.12 The branching raiding pattern of *E. burchelli* (*a*), and the trail produced in a computer model that takes account of the ants' trail-laying behaviour (*b*). (Images: Nigel Franks, University of Bristol.)

happen automatically if they merely have a different manoeuvrability when coming and going.

Aren't three lanes a little excessive, though? Surely two would do just as well for collision-avoidance; they work well on our roads. But the problem with two lanes is that the pattern is asymmetrical. Do outgoing ants stay to the right or the left? There's nothing to tell them which, unless the ants have some inbuilt mechanism to tell their right from their left. With three lanes, that problematic choice doesn't arise.

Not all ants that forage in trails form lanes like this. The leaf-cutter ants (*Atta cephalotes*) just push past each other. Why is that good enough for them and not for army ants? Perhaps part of the answer lies in the fact that they are in less of a hurry—they do not need to stop work at dusk, and so the selective pressure for efficiency is weaker. All the same, collisions surely make the task of gathering food less efficient than it might be. This, however, may not matter very much. Leaf-cutters carry huge burdens, and so the additional slowing they might experience from the lack of lanes may make little difference to their travel time—not enough, at least, to have made it worth their while acquiring the instincts that lead to lane formation. Some researchers have even suggested that collisions might not be all bad: they can, for example, help to transfer information between ants—the equivalent of a hiker coming in the other direction telling you that the cliff path ahead has collapsed.

Outside perfume advertisements, humans do not seem to follow each other's chemical trails. But there are other reasons why we tread in each other's footsteps. In deep snow we might literally do that, because it requires less effort and reduces the chances of plunging down a hidden hole. Another situation in which this trail-following may occur is on open grassy spaces. The ground may be smoother, for instance, where others have worn the grass away. And there is surely also a psychological impulse to 'stick to the path', even if we can see that that path has been defined by other walkers rather than stipulated by the planners. In this way, spontaneous paths are created and reinforced across the grass (Fig. 5.13a). If old paths are abandoned for some reason, the grass eventually grows back and covers them. This is entirely analogous to the way an ant pheromone trail gradually disperses and vanishes if not reinforced by others.

Dirk Helbing, Péter Molnár, and their colleague Joachim Keltsch in Tübingen wondered whether the pedestrian model might account for human trails of this sort. They introduced into the pedestrian model a tendency for walkers to walk where a trail has been worn down, this being itself determined by how many others had walked along that route. The path taken by a walker over an open space is therefore a compromise between this trail-following behaviour and the wish to take the most direct route. Unused trails became covered over at a steady rate.

The researchers found that at first walkers simply took straight-line routes between destinations (Fig. 5.13b). But, over time, the shapes of the trails altered: the straight lines disappeared, and instead curved routes emerged, with islands of grass isolated in the middle of intersections (Fig. 5.13c). These routes look less geometric and more 'organic'; but they are also less direct. They represent what the walkers 'perceive' to be the best trade-off between directness and ease of walking. Real human

FIG. 5.13 People walking over open grassy spaces wear down trails that have an 'organic' quality to them, with curving paths and smooth intersections (a). Here is one such, at Stuttgart University. In a computer model of trail-following pedestrian behaviour, similar trail patterns emerge over time (b, c). At first, the trails between entry and exit points (at the corners here) are linear and direct (b). But these evolve into curved trails that represent a compromise between directness and the tendency to follow in others' footsteps (c). (Photo and images: Dirk Helbing, Technical University of Dresden.)

trail systems seem to show the same characteristics (Fig. 5.13a). When the trails are all essentially going from one end of a space to the other, these spontaneous trails may fork and branch, sometimes creating routes that peter out (Fig. 5.14). These resemble the foraging trails made by hoofed animals through tall undergrowth (Fig. 5.14c).

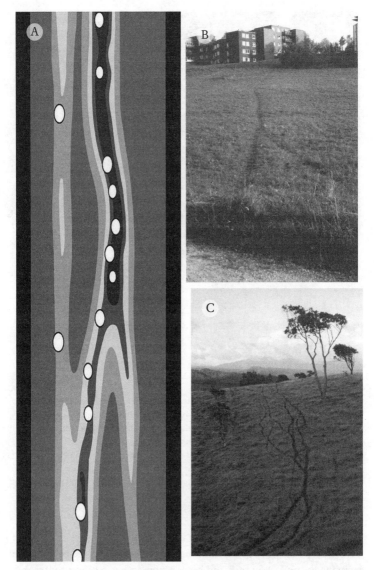

FIG. 5.14 For more linear motion, the trail-following model produces branching paths (*a*; here the white circles show some walkers on the paths), which resemble those seen for humans (*b*) and some foraging animals (*c*). (Images and photos: *a*, after Helbing *et al.* 1997; *b*, Dirk Helbing; *c*, Iain Couzin.)

HEAVY TRAFFIC

It doesn't take much imagination to see that Helbing's pedestrian model may have something to say about the flow of road traffic. On the road we are even more constrained in our choices, so there is even less room for free will to disrupt predictable, robotic behaviour. Road traffic is forced to move in a line along a pre-defined path, often in single file, and with an incentive to avoid collisions, motivated by strong concerns for our well-being and our bank balance. All we can really do is accelerate to our preferred cruising speed if we can, but slow down when there is a vehicle in our lane ahead.

Helbing and others have adapted the pedestrian model to this situation. You might expect that, with such a simple set of rules, the resulting behaviour would be also rather mundane. But these models generate traffic conditions every bit as rich, complex, and frustrating as those we experience daily behind the wheel. Again, the flow patterns show clear signs of being dominated by collective behaviour. For example, it seems reasonable to suspect that, as the traffic on a straight, one-lane road gets steadily denser, it will become progressively more congested and the average driving speed will gradually decrease until eventually a jam forms. But that is not what happens. Instead, there are rather abrupt switches between steady flow and almost immobile jams, once a critical threshold in traffic density (the number of vehicles per km of road, say) is exceeded. This is again analogous to a phase transition: a marked shift in 'global' behaviour brought about by only a small change in conditions. It is literally as though the traffic 'freezes', changing from liquid-like flow to solid-like stasis.

This sudden triggering of a jam is seen in a traffic model devised by Kai Nagel and Michael Schreckenberg, working in Germany in the early 1990s, which has more or less the rules indicated above.* Here a series of cars drives in procession down a straight road, all of them attaining the same speed if the road ahead is sufficiently clear. This means that on a graph of

*Another ingredient is a small amount of randomness in the speeding up and slowing down of each vehicle, to account for the fact that no one (whatever they might like to think) drives perfectly.

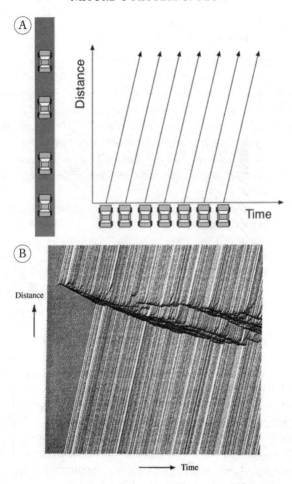

FIG. 5.15 Phantom jams in road traffic are a consequence of collective behaviour. The motion of a car moving down a road at a constant speed can be represented on a graph of distance travelled against time as a sloping straight line (a). In a computer model of many such cars moving in procession along a motorway, a single small disturbance caused by one vehicle slowing abruptly and briefly can develop into a complex jam (b, dark bands), which moves upstream and splits into several 'waves' of congestion. Observations of real road traffic have shown effects like this (c). (Images b and c: from Wolf, 1999.)

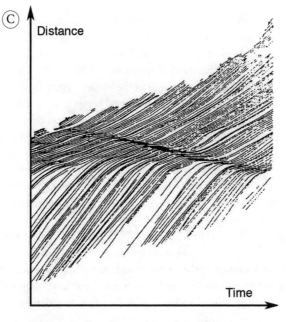

FIG. 5.15 (Continued).

distance travelled plotted against time, each car's trajectory traces out a straight, sloping line when it moves unimpeded (Fig. 5.15a). But if a car is forced to slow down, the line kinks towards the horizontal, since then the vehicle covers a shorter distance in the same time interval. When a single car is programmed to brake suddenly before speeding up again—mimicking, say, a driver whose attention wanders momentarily—the consequences are dramatic. This action forces the drivers immediately behind to brake as well, creating a little knot of congestion. This does not simply dissipate once the errant driver speeds up again, because the congestion makes others brake further upstream too, setting up a wave of congestion that propagates steadily in the opposite direction to the traffic flow (Fig. 5.15b). And as time passes, things go from bad to worse. This little jam keeps moving upstream, but it also gets wider and eventually forks in two, creating *two* knots of congestion. This branching continues, creating

153

a whole series of jams without any apparent 'cause'. A driver entering the scene long after this initial disturbance encounters waves of 'stop-and-go' traffic: a flow pattern that has acquired a momentum of its own.

Something like this has been observed in real road traffic (Fig. 5.15c). But here it doesn't look quite as bad as in the model, and indeed it turns out that the basic model of Nagel and Schreckenberg, while capturing the essential collectivity of drivers' behaviour, is a little too over-sensitive to minor disturbances. Boris Kerner and Hubert Rehborn of the Daimler-Benz research labs in Stuttgart think that congested traffic in fact has two states: a dense, almost immobile jam and a slightly less dense 'synchronized' state in which the traffic keeps flowing steadily at a moderate speed that is identical for all cars. Others maintain that the switch from free to synchronized flow has to be triggered by an external disturbance, such as a bottleneck on the road or a junction where another traffic stream joins.

Such disturbances are of course common on real roads, and may be responsible for much of the variation and complexity of real traffic. Dirk Helbing and his colleagues found that an entry junction onto a motorway, for example, can induce all kinds of traffic jams, from knots that move upstream or stay pinned to a point on the road, to oscillatory waves of congestion or solid block-like jams.

These models show that traffic flow has its own characteristic and robust patterns in time and space that emerge spontaneously and suddenly from the interaction of many individuals. This can make jams seem like a rather fundamental and inevitable part of life on the road; but the message need not be so gloomy. By helping us to understand under what conditions congestion and jams form, traffic models may enable us to lessen the chances of their happening, for example by appropriate design of road layouts, of speed restrictions, or of lane-changing rules. This might not only make the roads more pleasant, but could also make them safer, while reducing pollution. And models that can predict traffic flow on a real road network, based on measurements of traffic at a few key locations, could be valuable for planning a route or for alerting road authorities to potential congestion problems before they arise. Schemes like this, using models such as those I have described, are already being implemented in European and American cities and urban areas to allow real-time prediction of road use.

DON'T PANIC

The simulated crowds studied by Helbing and his colleagues are generally rather well-behaved, finding collective modes of movement that ease the flow and enable people to move past each other with the minimum of jostling and discomfort. When two such crowds converge at a doorway and try to pass through from opposite directions, they even display what looks like good manners, standing back from time to time for walkers coming the other way. (Of course, these 'manners' are illusory, since the simulated walkers have no scruples; their apparent thoughtfulness is prompted merely by the rule that avoids collisions.)

But not all crowds are so civilized. In crowd disasters, panic can lead people to push others over and trample them. Riots, building fires, and crushes at sports stadiums, rock concerts, and other public gatherings have in the past claimed many lives as a result of the uncontrolled movements of crowds that, in their terror or excitement, have ceased to show anything like the 'wisdom' with which they are fashionably attributed.

In 1999 Helbing teamed up with Tamás Vicsek and his colleague Illés Farkas in Budapest to try to understand what happens when a crowd panics. They studied how the 'model pedestrians' move when they want to get somewhere so quickly that the desire for speed overwhelms the impulse not to come into contact. The researchers found that a crowd like this can become jammed if all the individuals try to pass through a single doorway in a simulation of 'escape panic'. As walkers press in against one another in the dense throng in front of the door, they can become locked into arch-shaped lines, unable to move forward. It is precisely this sort of jamming that gives masonry arches their stability: the stones push against their neighbours to form a robust structure, held together by friction, which resists the tug of gravity. Such arches are also known to appear in grainy materials as they drop through a hole, which is why salt can get stuck in the salt cellar even though each of the individual grains is small enough to pass through the hole. Here, then, the 'crowd fluid' starts to act more like a 'crowd powder'. The result is counterintuitive in a rather chilling way: as everyone tries to move faster, the crowd actually exits through a doorway more slowly than it would if the individuals kept to a more moderate pace. The pressures that build up in the jammed crowd can be frightening: in real crowd disasters they have been large enough to bend steel bars and knock over walls.

Real animals do seem to behave this way. It is not the kind of experiment one can conduct ethically with human subjects, of course, and one winces even to think of it being done with mice. But Caesar Saloma and colleagues at the University of the Philippines, who did this in 2003, apparently caused no lasting harm to their subjects, which had to escape from a chamber that was being slowly flooded. They found that the mice showed just the kind of 'escape panic' exhibited by the computer model, and that the flow through the exit had the predicted feature of occurring in bursts of different sizes—very much like, in fact, the self-organized avalanches of the sand-pile model I described in the previous chapter.

Because of his work on pedestrian motion and particularly on crowd panic, Dirk Helbing was consulted in 2006 about crowd control for the annual pilgrimage of Muslims to Mecca in Saudi Arabia. This event, called the Hajj, draws up to four million pilgrims, and the threat of crowd disaster has been ever present. Hundreds of lives have been lost in the past—in 1990 over 1,000 people were killed in a stampede in a pedestrian tunnel leading out from Mecca to the nearby town of Mina, where a ritual stoning takes place.

This stoning has been a flashpoint for several crowd disasters. The pilgrims gather in Mina to cast stones at pillars called the jamarat, re-enacting the stoning of the devil by Abraham. To ease congestion, the three pillars have been replaced by elongated, elliptical structures more like walls, and a two-tiered 'bridge' has been built so that more pilgrims can gain access to the jamarat at the same time. But these measures have proved inadequate: pilgrims were trampled to death during the ritual on six occasions since 1994. The disaster in January 2006 was one of the worst: over 300 pilgrims died, and many more were injured. The Saudi authorities organizing the Hajj realized that something had to be done. Could Helbing's crowd model help?

For the 2006 event, the authorities had installed video cameras to monitor the movements of the crowd. They allowed Helbing and his colleagues to study the footage, hoping that this might reveal how the crush became fatal. What the researchers saw surprised and shocked them.

As the crowds on the Jamarat Bridge became denser, the steady flow changed to stop-and-go waves like those in heavy traffic (Fig. 5.16a)—something that had not been so clearly seen before in crowd motion. But then, as the throng thickened still further, this motion, which was uncomfortable and frustrating but nonetheless relatively orderly, gave way to another kind. People became clustered into knots, reminiscent of the eddies of fluid flow, which swirled around in all directions (Fig. 5.16b). Pilgrims were pushed here and there, powerless to do anything about it, with enough force to knock people off their feet. If they stumbled, they might have no chance to get up again, nor could their neighbours avoid trampling on them. The movement looked startlingly like turbulence in a rushing liquid.

This 'crowd turbulence' is not predicted by the simple models of pedestrian motion or crowd panic. Helbing and his colleagues think that it is triggered when individuals alter their behaviour in response to the crush: instead of being passively pushed along by the crowd, people try to ease the squeezing they experience by pushing back. The 'crowd fluid' becomes animated, spontaneously injecting more energy into the motion. That is something a simple fluid cannot do, and so it is no surprise that 'crowd turbulence' doesn't exactly mirror ordinary turbulence. It does mean, however, that the situation suddenly becomes much more violent and dangerous.

The researchers deduced that the onset of this hazardous 'turbulence' has a special signature that can be spotted in video recordings. It happens at a threshold determined not by crowd density alone but also by how much variation there is in the speed of individuals' motions. Together, these two factors define a critical 'crowd pressure' at which turbulence sets in. So real-time monitoring and analysis of video data on dense crowds can give advance warning of when this highly dangerous state is about to develop, potentially allowing stewards to introduce crowd-control measures (such as opening up new exits or stopping further influx) that might relieve the pressure and avoid a fatal incident.

That, however, happily did not become necessary in the 2007 Hajj. As a result of the studies by Helbing's team, a new route between the pilgrim camp and the jamarah at Mina was devised, with specific streets designated for one-way flows and stringent schedules arranged to limit and

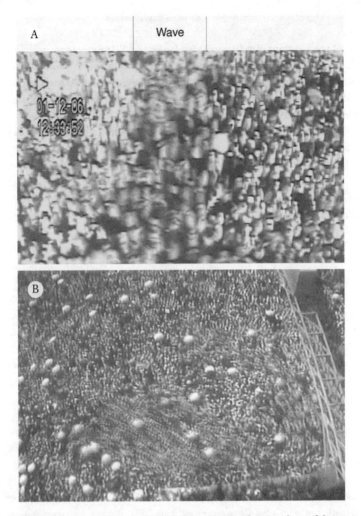

A   Wave

01-12-06
12:39:52

B

FIG. 5.16 Crowd movements at the 2006 Hajj in Mina. Video recordings of the event revealed various modes of motion, including stop-and-go-waves (a) and crowd 'turbulence' (b). Both images were made by averaging the video frames over about 1–2 seconds, so that stationary people appear sharp while moving ones are blurred. The stop-and-go waves in a propagate from left to right, while the crowd is moving from right to left. This image was taken on a street leading to the Jamarat Plaza. Crowd turbulence (b) is characterized by waves (clusters of walkers) moving in all directions. This figure is produced from a video at the entrance to the Jamarat Bridge; one of the ramps leading onto the bridge is visible on the right. (Photos: Dirk Helbing and Anders Johansson, Technical University of Dresden.)

distribute the flow of pilgrims. Despite the fact that even more pilgrims came to the Hajj than were expected, the plan proved a complete success, and the rituals passed without incident. There could hardly be a better testament to the value of understanding the patterns of the crowd.

INTO THE MAELSTROM

The Trouble with Turbulence

W hat would you ask God? The German physicist Werner Heisenberg allegedly had this in mind: 'When I meet God, I am going to ask him two questions: why relativity? And why turbulence? I really believe he will have an answer for the first.'

The quote is apparently apocryphal, although not implausible: turbulence was the topic of Heisenberg's doctoral thesis in 1923. But, like most such stories, it was coined to make a point: understanding turbulent fluid flow is so hard that neither Heisenberg nor, perhaps, God could achieve it. Heisenberg's name might have become linked to the tale simply because he was better known than Sir Horace Lamb, a British mathematician and an expert on fluid mechanics, who does seem to have said something similar in 1932 in an address to the British Association for the Advancement of Science.*

Whether it is genuine or not, Heisenberg's remark is illuminating, because it illustrates different ways in which a scientific problem can be perplexing. One is that the phenomena themselves lie outside our mundane experience. The theory of relativity developed by Albert Einstein might, at least from the perspective of the 1930s, be deemed a piece of deistic obfuscation because it is well hidden in our everyday world and therefore

---

*Lamb's comment was reportedly 'I am an old man now, and when I die and go to heaven there are two matters on which I hope for enlightenment. One is quantum electrodynamics, and the other is the turbulent motion of fluids. And about the former I am rather optimistic.'

seems rather superfluous. It is not immediately clear why any deity would feel the need to build a universe governed by anything other than the old Newtonian laws of mechanics. Why dictate that these must give way to strange relativistic effects such as the shrinking of space and the stretching of time when objects move very fast? The maths needed to comprehend relativity is hard, but not burdensomely so by the standards of theoretical physics. Yet the concepts involved defy our experience and intuitions.*

Other problems in science may be difficult because they really do demand a degree of mathematical abstraction and sophistication that is not accessible to most of us. Superstring theory is somewhat like this: most of us can see that a lifetime would not be enough to decode its equations.

But turbulence is difficult and perplexing in yet another way. The basic question is simple: how do we describe a fast-flowing fluid in mathematical terms? When flow is vigorous enough, the regular structures and patterns that we have seen in earlier chapters tend to dissolve, leaving an apparent chaos that changes with every passing moment (Fig. 6.1). And yet this does *not* destroy all structure, for then the liquid would be pervaded by random motion throughout, so that on average the flow is merely uniform. We *do* see patterns in turbulent flow, as Jean Leray surely appreciated in the early twentieth century while gazing at the eddies of the Seine. Swirling vortices are continuously born and swallowed up, offering tantalizing hints of order. But how do we capture and describe that order?

It is not that we don't know how to build a theory—the principles governing the flow are actually remarkably straightforward. We simply apply Newton's laws of motion throughout the fluid, which describe how the fluid's velocity changes in proportion to the forces that act on it. The problem is that we can't solve these equations. They are too complicated, because the fluid motion is now totally interdependent: the movement of each little 'piece' of fluid depends strongly on that of all the surrounding

---

*I do not mean to imply that a Newtonian universe would 'work'; we simply don't know if it would or not. Of course, in reality relativity does not supplant Newton's laws but *explains* them as a special case that applies at slow speeds and in moderate gravity. But Heisenberg was understandably puzzled about why there should be any such distinctions. Many physicists hope that one day a unified theory will show why relativity is an essential aspect of the way things are.

FIG. 6.1 In turbulent flow, the motion of the fluid is chaotic, and yet some coherent structures such as vortices survive. (Photo: Katepalli Sreenivasan, Yale University.)

pieces. In a sense, this is always true in a fluid; but when it becomes turbulent, it is no longer possible to make approximations or to take averages. Every detail matters. So the problem is difficult not because we don't know what the ingredients are, but because those ingredients are too mixed up to make sense of them. There is just too much going on.

Many of the greatest scientists have bloodied their knuckles against the implacable walls that surround the problem of turbulent fluid flow. David Ruelle, a physicist who has contributed a great deal to our understanding of it, has called turbulence 'the graveyard of theories'. He notes with glee how the classic book *Fluid Mechanics* by the Russian physicists Lev Landau and Evgeny Lifshitz, which generally displays the characteristically uncompromising attitude to mathematical exposition common in the Russian literature, reverts to pure narrative description when the authors come to talk about turbulence. The equations no longer help. These formidable scientists were instead forced to do what Chinese artists have long done: to use pictures.

This is, however, perhaps to paint too bleak a portrait of where our understanding of the patterns of turbulent flow has reached today. We do know a great deal about these patterns, and we can say some important things about the recondite 'geometry' of turbulence. This is what I hope to give you a flavour of now.

THE MASTER EQUATION

Isaac Newton's *Principia* provides a prescription for how bodies move: they change their velocity (that is, they accelerate) when a force acts on them. The rate of change of velocity is equal to the force divided by the object's mass: this is Newton's celebrated second law of motion. In the middle of the nineteenth century an Irish mathematical physicist named George Gabriel Stokes wrote down an equation for fluid motion based on Newton's second law. This equation was really just a more rigorous restatement of a formula derived by the French engineer Claude-Louis Navier in 1821, and so it bears the name of the Navier–Stokes equation. It says that the rate of change of velocity at all points in a fluid is proportional to the sum of the forces that promote its movement, such as pressure and gravity, along with the retarding force of viscous drag exerted by the surrounding fluid. The Navier–Stokes equation is a little set of equations that specify Newton's second law, and to fully describe the flow they are supplemented by other equations specifying that mass and energy be conserved (nothing gets lost) as the flow proceeds.

The catch, as I say, is that the Navier–Stokes equations are often exceedingly difficult to solve without making assumptions and approximations about the behaviour of the fluid. In effect, solving the equations requires that you already know the answer—for you can only calculate the viscous drag on a 'parcel' of fluid if you know what all the surrounding parcels are doing, and yet the same problem applies to those too. Much of the theoretical work on fluid dynamics today revolves around the issue of how to introduce appropriate simplifications into these equations describing fluid flow for particular types of flow so that they can be solved without in the process losing the essential features.

Lord Rayleigh did precisely this when he developed a theory of convective flow in the early twentieth century (see page 51). We saw earlier that he showed how convection patterns (ordered arrays of 'cells' with circulating flows) arise above a particular threshold in the strength of the driving force, namely the temperature difference between the cool top and hot bottom of the layer of fluid. If this forcing is very strong, convection becomes turbulent. The switch from regular patterned flow to turbulence is not gradual, but rather abrupt. Yet as we saw on the surface of the Sun, the onset of turbulence does not mean that the flow loses all structure; it's just that the structure changes over time in unpredictable ways.

Osborne Reynolds found the same thing for fluid flow along a cylindrical pipe. In 1883, he showed that there is a transition from smooth, eddy-free (so-called laminar) flow to turbulent flow down pipes as the flow rate (quantified by the Reynolds number) increases. Reynolds also sought to understand this switch using the Navier–Stokes equations. The equations in this case cannot be solved with pen and paper, but they can be solved numerically on a computer: the patterns of fluid flow that satisfy the equations are found by a series of iterations that refine some rough initial guess. This reveals an odd thing, however. The computer calculations do not seem to show a turbulent threshold at all; instead, the flow may remain laminar for all Reynolds numbers.

And yet, in practice, most pipe flows with a Reynolds number above about 2,000 are turbulent. This is a typical value for water running from a tap, which does indeed come out as a turbulent jet (Fig. 6.2). Why is there this discrepancy between theory and experiment? It seems that the transition to turbulence depends on the flow being disturbed: there has to be a 'kick' to trigger the appearance of eddies. In experiments in which the flow down a pipe is carefully controlled to avoid such disturbances, laminar flow seems to persist at least up to Reynolds numbers of about 100,000. The bigger the Reynolds number (the faster the flow), the smaller the kick required—the flow is more delicately poised to become unstable, as we might expect.

To further complicate matters, turbulence in pipe flow does not last for ever. If triggered by some disturbance, the turbulence seems to persist only for a certain time before the flow becomes smooth (laminar) again. Equivalently, turbulence caused by a persistent perturbation at one point in the pipe (a bump in the wall, say) will wash out eventually further down the pipe, if it is long enough. This settling might take a very long time, however. It is estimated that you would need to wait five years for the flow to travel 40,000 km along a garden hose before turbulence excited in one segment decays again.

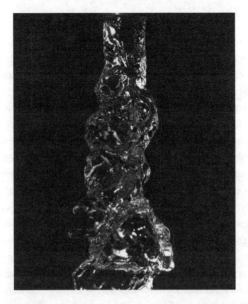

FIG. **6.2** Water emerges from a tap as a turbulent jet, here captured in freeze-frame by the high-speed photography of Harold Edgerton. (Photo: The Edgerton Center, Massachusetts Institute of Technology.)

So even if we can solve the Navier–Stokes equations by the brute force of computer number-crunching, such equations may not tell us all we need to know, because they don't take into account how the flow might respond to the kind of perturbation that, in the real world, it is very likely to encounter. The question is whether or not the resulting disturbance of the flow will gradually die away, and if so, how quickly. That is an important distinction, because the difference between smooth and turbulent flow can be crucial in an industrial context. In turbulent flow, the fluid becomes strongly mixed up. And turbulence in pipe flow can hinder the passage of the fluid: the eddies, you might say, get in the way, reducing the overall flow rate. This may cause pressure surges in the pipe, which matters a great deal if we're talking about oil, gas or water flowing down service pipes, or chemical liquids being carried between vats and tanks in a chemical processing plant. The issue might be even more vital, so to speak, in the case of blood flowing through the pipelines of the body's circulatory system—a topic I consider in Book III.

CANNED ROLLS

The question of how smooth and ordered flows capitulate to turbulence has therefore been studied a great deal. Pipe flow and convection provide two convenient experimental settings for these studies, but there is a third that offers another attractive demonstration of how fluid flow tends to progress through a series of regular patterned states before arriving at turbulence. In 1888 the French fluid dynamicist Maurice Couette looked at flow induced in a fluid sandwiched in the gap between two concentric cylinders of different sizes. To induce flow, the inner cylinder was rotated, which drags the fluid next to the wall along with it (Fig. 6.3a). This is now called Couette flow.

In some ways this is similar to flow down a parallel-sided channel, which we considered in Chapter 2: at low rotation speeds, the fluid's velocity changes smoothly across the flow profile, so that the fluid can be regarded as a series of thin layers sliding past one another (Fig. 6.3b). (This is precisely why smooth flows of this sort are said to be laminar, as in a laminate.) But one key difference is that a rotating object experiences a

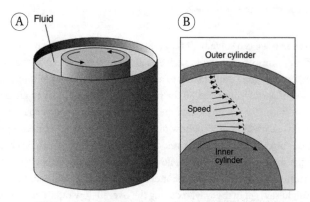

FIG. 6.3 In the apparatus devised by Maurice Couette, a fluid is held within two concentric cylindrical drums, and is set in motion by rotating the inner drum (a). Geoffrey Taylor later modified the device such that the outer drum might be rotated too. The fluid is dragged along by friction at the interface with the inner cylinder, and can be considered to move in a series of concentric shells with different velocities—a shear flow (b).

centrifugal force—the force that pulls tight the string when a threaded weight is spun in a circle. So not only is the fluid carried around in circles, but it is simultaneously forced outwards. As ever, viscous drag resists this outwards force, so that for low rotation speeds the centrifugal force does not appear to affect the flow.

But the British mathematician Geoffrey Taylor found in 1923 that, once the centrifugal force starts to overwhelm the damping effect of viscosity, patterns appear. First, the column of fluid develops stripes (Fig. 6.4a).

FIG. 6.4 Various pat-terns form in Couette flow as the rotation speed of the inner drum (and thus the Reynolds number) increases. First, a stack of doughnut-shaped roll cells is formed (a). These then develop wavy undulations (b). At higher Reynolds number, the roll cells persist but each of them contains turbulent fluid (c). Finally, the fluid becomes fully turbulent (d). But even here, the turbulence may be intermittent in time and space: there is a region of smooth flow here amidst the turbulence. (Photos: from Tritton, 1988.)

These are in fact roll-like vortices in which the fluid circulates in alternate directions, as if around the surfaces of a stack of doughnuts. Like Rayleigh–Bénard convection, this is a symmetry-breaking process that creates a pattern of a well-defined size.

It isn't too hard to see that this situation is closely analogous to convection, in which the same kind of symmetry-breaking structure (roll cells) is created. All the fluid in the inner part of the Couette flow 'tries' simultaneously to move outwards because of the centrifugal force. But it cannot simply pass through the outer layers. At a threshold rotation speed, the system becomes unstable so that roll vortices transport part of the inner fluid to the outer edge while a return flow replenishes the inner layer. Not only is the instability of the same basic nature as that in convection, but the shape of the rolls is the same: roughly square, as wide as the gap between the inner and outer cylinders.

The 'dimensionless number' that characterizes the driving force for flow is again the Reynolds number. Here it is defined according to the velocity of the flow at the surface of the inner cylinder, while the characteristic dimension of the system is the width of the gap between the two cylinders. Taylor performed a calculation analogous to Rayleigh's for convection, to work out when the roll cells—now known as Taylor vortices—will appear as the Reynolds number is increased.

Increasing this driving force further by spinning the apparatus more rapidly produces a progressive elaboration of the basic stripy pattern. First the roll cells become wavy, undulating up and down around the cylinders (Fig. 6.4b); then the waves get more complex before becoming more or less turbulent—after which the stacked stripes reappear with turbulence inside them (Fig. 6.4c). Finally, when the Reynolds number is about 1,000 times the value at which the pattern first appears, the whole column of fluid becomes an unstructured wall of turbulence (Fig. 6.4d).

But there is more. Taylor realized that the game changes if, instead of keeping the outer cylinder fixed, we let that rotate as well. Then the fluid can experience significant centrifugal forces even when the *relative* rotation speed of the inner layer with respect to the outer is small, and so a different balance of forces can be established. Experiments on a system like this have revealed a menagerie of patterns, too numerous to show here: interpenetrating spirals, wavy vortices, corkscrew wavelets, spiral turbulence. As we have seen, some of these flow patterns may be analogous to those found in a rotating planetary atmosphere.

HIDDEN ORDER

If you drive a fluid flow hard enough you will always end up with turbulence—with chaotic flows that constantly shift. But there are several ways in which the transition to turbulence may come about. For convection, the switch tends to be abrupt. In the wakes of shear flows, such as fluid flowing around an obstacle, turbulence first comes and goes intermittently, and only takes hold fully when the flow is faster. In Taylor–Couette flow, turbulence and regular patterns can coexist for a while in the form of turbulent Taylor vortices. There are many routes to turbulence, and those taken by particular types of flow are still being debated.

When turbulence does finally arrive in earnest, however, we might be tempted to give up looking for any pattern in it all. The trajectories of the fluid particles become extremely convoluted and ephemeral, and the Navier–Stokes equations can be solved only by laborious computer calculations, not by mathematical ingenuity. A turbulent fluid is in a state of continual instability: you could say that every single thing that happens in the flow is catastrophic, perturbing everything else. This means that we generally cannot predict anything about how the flow will evolve or where any particles carried within it will be at any point in time. (It does not mean, however, that the Navier–Stokes equations break down, but instead that these equations no longer have solutions that remain unchanging in time.)

In this situation, rather than trying to look at the detailed pattern of flow in terms of streamlines, we are better off just asking about average features—in other words, we must forget about individual trajectories of fluid particles and consider instead their statistical properties. Then, even apparently random, structureless systems like turbulent fluids prove to have characteristic forms, just as we found a kind of non-random 'order' emerging from the self-organized landslides of sand grains in Chapter 4. In this way we might be able to distinguish one kind of apparently chaotic process from another by comparing their statistical forms. We will see in Book III some further illustrations of this important concept of 'statistical form'.

The idea that turbulent flows have a generic statistical 'shape' which can be measured quantitatively has been explored throughout the last century. In the 1920s the Englishman Lewis Fry Richardson proposed that the 'universal' properties of turbulence should become apparent if we divide it up into the average 'global' fluid velocity and the deviations from that average (the fluctuations) at each point. We can think of a fluid as having a particular flow velocity (speed and direction) at every point (a 'velocity field'), as though each little parcel of fluid were like one of the swarming particles we encountered in the previous chapter. Most turbulent flows have non-zero mean velocities: the fluid does 'get somewhere', albeit in a haphazard fashion. Think, for example, of the turbulent wake in river flow past a pillar, or a turbulent jet of smoke exhaled by an office worker as he stands fashionably exiled in the cold outdoors. Richardson suggested that the generic behaviour of turbulence resides in the statistics of the fluctuations, from which any average flow must first be subtracted.

He proposed that any structure buried in the fluctuating, chaotic part of the velocity field could be revealed by considering how the differences in velocity at two points within the flow vary as the points get farther apart. There is no single answer to this: it's a matter of collecting statistics. If the flow is totally random at all scales, the velocity at one point will bear no relation to that at any other: all differences in velocity will occur with equal probability as the points get further apart. If the flow has a structure, however, like a convection cell, the velocities at different points will tend to be related to one another in some non-random way, so that knowing one allows us to predict (or at least estimate) the other. In such a case, the velocities are said to be *correlated*. For example, the velocities in adjacent edges of two roll cells are not independent: if the fluid at a point on one edge is going upwards, we can be sure that the fluid at a corresponding point on the edge of the other cell is also moving upwards at about the same speed, because adjacent rolls are always counter-rotating.

In economics, some market traders spend a lot of time looking for correlations between stock prices, so that one may be forecast from another, or so that the price at some point in the future can be deduced from that today. It seems clear that correlations in time disappear rather quickly in stock price fluctuations—but if they can be glimpsed, there may be money to be made if you are swift enough. As it happens, it has been suggested (albeit controversially) that price correlations in economics may show some similarity to those of turbulent fluid flow—in which case, talk of 'market turbulence' is not purely metaphorical.

If turbulence has inherent structures that distinguish it from utter randomness, there will be some correlation between the velocities at different points in the flow. Intuitively we should expect that in a chaotic flow such correlations, if they exist at all, will be less strong the further apart the points are. Notice that in a perfectly ordered array of Rayleigh–Bénard convection cells this is not the case: the correlations are very long-ranged, since the cells are arranged in an orderly manner. Yet experiments have shown not only that there *are* correlations in turbulence, but that these have a remarkably long reach, generally extending over almost the entire width of the flow. It is as though individuals in a jabbering crowd were able to converse with one another from opposite sides of a room.

These correlations make a description of turbulence a subtle business. They are responsible for its elegant, baroque beauty, a confection of swirling, vortex-like structures of many different sizes. Fully fledged turbulence is often patchy, in fact resembling a sluggish river, in which regions of intense disorder and 'swirliness' are superimposed on a more quiescent background. We saw in Chapter 2 that one of the fundamental structures of turbulence is the whirlpool-like vortex or eddy. But whereas in pre-turbulent flow the eddies may form highly regular patterns such as Kármán vortex streets, turbulent eddies are formed over a very wide range of size scales, and they are transient (like Jupiter's Great Red Spot) and might appear anywhere in the flow.

Eddies carry much of the energy of a turbulent flow. Whereas in a laminar flow the energy is borne along in the direction of the fluid motion, in turbulent flow only a part of the motional (kinetic) energy of the fluid gets anywhere—the rest is captured by eddies, which fritter it away, in the end dissipating it in frictional heating as one parcel of fluid rubs against another. (This friction is the origin of viscosity.) The dissipation of kinetic energy happens at very small length scales, as the molecules in the fluid collide and increase their thermal jiggling. So the energy that is fed into the flow at large scales, creating big eddies that we can see, finds its way by degrees down to these small scales before being dissipated. In other words, there is an *energy cascade*: big eddies transfer their energy to smaller eddies, which do likewise at ever smaller scales. Richardson appreciated this, and in 1922 he coined a rhyme, inspired by Jonathon Swift's doggerel about fleas, to describe the process:

> Big whirls have little whirls
> that feed on their velocity,
> and little whirls have lesser whirls
> and so on to viscosity.

In the 1940s the Russian physicist Andrei Kolmogorov put this energy cascade into precise mathematical form. He proposed that the energy contained in a turbulent fluid at a length scale d varies in proportion to the 5/3rd power of d—in other words, it increases as d gets larger at a rate proportional to slightly less than the square of d. Compare this with how the area of a circle of diameter d increases with d: in that case, it does so at a rate proportional to d squared.* This is another example of a power law (see Chapter 4), also called a *scaling law* because it describes how some quantity varies with a change of scale. Scaling laws are central to the science that underlies many natural patterns and forms—we will encounter others in Book III.

Kolmogorov's law no doubt sounds rather abstract, but what it tells us is that there is a kind of logic to the process of turbulent flow which governs the way energy gets distributed in the fluid. It turns out that his scaling law is often found to be slightly inaccurate when investigated experimentally, since Kolmogorov made slightly too simplistic an assumption in deriving it. But more recent theories of turbulence have shown that things can be put right by including a few other ingredients in the scaling law. The basic idea of an energy cascade, in which the energy magnification in the fluid flow at different scales is described by a power law, is correct.

It is not obvious what this law implies for the way turbulence *looks*. There is, however, a rather delightful way of illustrating this. We have seen already that one of the characteristic forms of turbulent flow is the

---

*If you increase the diameter of a circle by a factor of 10, you increase its area by a factor of 10^2 = 100. But if you compare a patch of turbulent fluid with another one ten times the width, the energy in the latter is greater by a factor of $10^{5/3}$, or about 47.

vortex or eddy. When, in 2004, scientists using the Hubble Space Telescope saw such features in a turbulent, expanding cloud of dust and gas around a distant star (Fig. 6.5*a*), it reminded them of Vincent van Gogh's famous painting *Starry Night* (Fig. 6.5*b*), which he completed in a mental asylum at St-Remy in 1889, a year before his death. Such comparisons prompted a team of scientists in Spain, Mexico, and England to

FIG. **6.5** Turbulence seen in interstellar gas and dust around the star V838 Monocerotis (V838 Mon), about 20,000 light-years away from Earth in the direction of the constellation Monoceros (*a*). The image was taken by the Hubble Space Telescope in February 2004. The dust is illuminated by a flash of light emitted from V383 Mon, a red supergiant star, at the middle of the picture. This image made astronomers think of Van Gogh's famous painting *Starry Night* (1889) (*b*). (Photos: *a*, NASA, the Hubble Heritage Team (AURA/STScI) and ESA. *b*, Digital image copyright 2008, Museum of Modern Art/Scala, Florence.)

FIG. 6.5 (Continued)

wonder whether van Gogh's trademark swirly style is genuinely turbulent in a scientific sense. To assess this, they looked at whether the statistical distribution of brightness variations in the painting has the form that Kolmogorov specified for turbulent flow.

From digitized images, the researchers measured the statistics of how the brightness varied between any two pixels a certain distance apart. They reasoned that this distribution can be considered analogous to the variations in velocity of fluid parcels in a turbulent flow—which, according to Kolmogorov's theory, should obey particular power laws.* For *Starry Night*, these correspondences hold fast with impressive accuracy. In other words, the painting gives a technically accurate representation of what Kolmogorov's turbulence 'looks like'.

---

*To be specific: Kolmogorov's work led to the prediction that the distribution of velocity differences between two points, δv, obeys a series of different power laws corresponding to different powers of δv: there is one law for δv^2, one for δv^3, and so on. The researchers searched for similar power-law relationships governing the differences in brightness in Van Gogh's painting.

The same is true for Van Gogh's *Road with Cypress and Star* and *Wheat Field with Crows*, both painted in a particularly disturbed period in early 1890 (the second of these paintings was the last he completed before he shot himself). But his notorious *Self-portrait with Pipe and Bandaged Ear* (1888) doesn't show the imprint of 'turbulence'. Could this be because it was executed while in a self-confessed state of calm, after he had been hospitalized and treated with potassium bromide for his psychosis? It is perhaps rather fanciful to imagine that Van Gogh's mental 'turbulence' gave him the ability to intuit the forms of real turbulent flow. But, whatever the reason, he was clearly able to do so on some occasions, and it may be for this reason that we can sense so strongly the discord, the constant and always imminent dissolution of order into chaos, in an image like *Starry Night*. We have seen this wild pattern before.

BÉNARD CONVECTION

P olygonal convection cells will appear in a thin layer of a viscous liquid heated gently from below. This is a classic 'kitchen' experiment, since it involves little more than heating oil in a saucepan on a cooker. The base of the pan must be flat and smooth, however, and preferably also thick to distribute the heat evenly. A skillet works well. The oil layer need be only about 1–2 mm deep. The flow pattern can be revealed by sprinkling a powdered spice such as cinnamon on to the surface of the oil.

For a more controlled experiment, silicone oil works better. This is commercially available in a range of viscosities, and a viscosity of 0.5 cm^2/ s is generally about right. The convection cells can be seen more clearly if metal powder is suspended in the fluid (see Fig. 3.1). Bronze powder can be obtained from hardware shops or art material suppliers. Aluminium flakes can be extracted from the pigment of 'silver' model paints by decanting the liquid and then washing the residual flakes in acetone (nail-varnish remover). These powders will settle in silicone oil if left to stand.

These procedures are based on:

S. J. VanHook and M. Schatz, 'Simple demonstrations of pattern formation', *The Physics Teacher* 35 (1997): 391.

This paper provides the names and addresses of some US suppliers of the substances required.

GRAIN STRATIFICATION
IN A MAKSE CELL

T his is one of the most satisfying experiments, giving a dramatic and dependable result for rather little effort. I have used it in several demonstration lectures—it is portable and reusable, and always elicits a satisfying response. I understand that the discoverers of the effect at Boston University have made a cell 2 feet high for such demonstrations. My Makse cell is no masterpiece of engineering, but it is quick and easy to make. It is convenient to make the transparent plates detachable so that they can be cleaned. Ideally they should also be treated with an anti-static agent, such as those used on vinyl records, to prevent grains from sticking to the surface—but this isn't essential. The plates are 20 by

30 cm, with a gap of 5 mm between them. The cells described in the original paper by Makse *et al.* (1997) were left open at one end, but an endpiece at both ends ensures that the plates remain parallel and means that the layers can fill up the cell completely, which gives a more attractive and easily visible effect. The prettiest results are achieved with coloured sand grains, which can be bought from some chemical suppliers. But granulated sugar and ordinary sand (cleaned, from a pet shop or from children's sand pits) are easier to get hold of, and have enough of a difference in grain size, shape and colour to produce visible layering. The sugar grains are larger and more square—table salt is too similar to sand, and so it doesn't segregate well. The best results are obtained by pouring a 50:50 mixture of grains at a slow and steady rate into one corner of the cell. For a funnel, all you need is an envelope of about A5 size with the tip of a corner cut off.

BIBLIOGRAPHY

Anderson, R. S., 'The attraction of sand dunes', *Nature* 379 (1996): 24.

Anderson, R. S., and Bunas, K. L., 'Grain size segregation and stratigraphy in Aeolian ripples modeled with a cellular automaton', *Nature* 365 (1993): 740.

Aragón, J. L., Naumis, G. G., Bai, M., Torres, M., and Maini, P. K., 'Kolmogorov scaling in impassioned van Gogh paintings', *Journal of Mathematical Imaging and Vision* 30 (2008): 275.

Bagnold, R. A., *The Physics of Blown Sand and Desert Dunes* (London: Methuen, 1941).

Bak, P., *How Nature Works* (Oxford: Oxford University Press, 1997).

Bak, P., Tang, C., and Wiesenfeld, K., 'Self-organized criticality. An explanation of $1/f$ noise', *Physical Review Letters* 59 (1987): 381.

Bak, P., and Paczuski, M., 'Why Nature is complex', *Physics World* (December 1993): 39.

Ball, P., *Critical Mass* (London: Heinemann, 2004).

Barrow, J. D., *The Artful Universe* (London, Penguin, 1995).

Ben-Jacob, E., Cohen, I., and Levine, H., 'Cooperative self-organization of microorganisms', *Advances in Physics* 49 (2000): 395.

Buhl, J., Sumpter, D. J. T., Couzin, I. D., Hale, J. J., Despland, E., Miller, E. R., and Simpson, S. J., 'From disorder to order in marching locusts', *Science* 312 (2006): 1402.

Camazine, S., Deneubourg, J.-L., Franks, N. R., Sneyd, J., Theraulaz, G., and Bonabeau, E., *Self-Organization in Biological Systems* (Princeton: Princeton University Press, 2001).

Cannell, D. S., and Meyer, C. W., 'Introduction to convection', in Stanley, H. E., and Ostrowsky, N. (eds), *Random Fluctuations and Patterns Growth* (Dordrecht: Kluwer, 1988).

Couzin, I. D., and Franks, N. R., 'Self-organized lane formation and optimized traffic flow in army ants', *Proceedings of the Royal Society London B* 270 (2002): 139.

Couzin, I. D., Krause, J., James, R., Ruxton, G. D., and Franks, N. R., 'Collective memory and spatial sorting in animal groups', *Journal of Theoretical Biology* 218 (2002): 1.

Couzin, I. D., and Krause, J., 'Self-organization and collective behavior in vertebrates', *Advances in the Study of Behavior* 32 (2003): 1.

Couzin, I. D., Krause, J., Franks, N. R., and Levin, S. A. 'Effective leadership and decision-making in animal groups on the move', *Nature* 433 (2005): 513.

Czirók, A., and Vicsek, T., 'Collective behavior of interacting self-propelled particles', *Physica A* 281 (2000): 17.

Czirók, A., Stanley, H. E., and Vicsek, T., 'Spontaneously ordered motion of self-propelled particles', *Journal of Physics A: Mathematical and General* 30 (1997): 1375.

Czirók, A., and Vicsek, T., 'Collective motion', in Reguera, D., Rubi, M., and Vilar, J. (eds), *Statistical Mechanics of Biocomplexity, Lecture Notes in Physics* 527 (Berlin: Springer-Verlag, 1999): 152.

Durán, O., Schwämmle, V., and Herrmann, H., 'Breeding and solitary wave behavior of dunes', *Physical Review E* 72 (2005): 021308.

Endo, N., Taniguchi, K., and Katsuki, A., 'Observation of the whole process of interaction between barchans by flume experiment', *Geophysical Research Letters* 31 (2004): L12503.

Forrest, S. B., and Haff, P. K., 'Mechanics of wind ripple stratigraphy', *Science* 255 (1992): 1240.

Frette, V., Christensen, K., Malthe-Sørenssen, A., Feder, J., Jøssang, T., and Meakin, P., 'Avalanche dynamics in a pile of rice', *Nature* 379 (1996): 49.

Glatzmaier, G. A., and Schubert, G., 'Three-dimensional spherical models of layered and whole mantle convection', *Journal of Geophysical Research* 98 (B12) (1993): 21969.

Gollub, J. P., 'Spirals and chaos', *Nature* 367 (1994): 318.

Grossmann, S., 'The onset of shear flow turbulence', *Reviews of Modern Physics* 72 (2000): 603.

Helbing, D., Keltsch, J., and Molnár, P., 'Modelling the evolution of human trail systems', *Nature* 388 (1997): 47.

Helbing, D., Farkas, I., and Vicsek, T., 'Simulating dynamical features of escape panic', *Nature* 407 (2000): 487.

Helbing, D., Molnár, P., Farkas, I. J., and Bolay, K. 'Self-organizing pedestrian movement', *Environment and Planning B: Planning and Design* 28 (2001): 361.

Helbing, D., 'Traffic and related self-driven many-particle systems', *Reviews of Modern Physics* 73 (2001): 1067.

Helbing, D., Johansson, A., and Al-Abideen, H. Z., 'The dynamics of crowd disasters: an empirical study', *Physical Review E* 75 (2007): 046109.

Henderson, L. F., 'The statistics of crowd fluids', *Nature* 229 (1971): 381.

Hersen, P., Douady, S., and Andreotti, B., 'Relevant length scale of barchan dunes', *Physical Review Letters* 89 (2002): 264301.

Hill, R. J. A., and Eaves, L., 'Nonaxisymmetric shapes of a magnetically levitated and spinning water droplet', *Physical Review Letters* 101 (2008): 234501.

Hof, B., Westerweel, J., Schneider, T. M., and Eckhardt, B., 'Finite lifetime of turbulence in shear flows', *Nature* 443 (2006): 59.

Houseman, G., 'The dependence of convection planform on mode of heating', *Nature* 332 (1988): 346.

Ingham, C. J., and Ben-Jacob, E., 'Swarming and complex pattern formation in *Paenibacillus vortex* studied by imaging and tracking cells', *BMC Microbiology* 8 (2008): 36.

Jaeger, H. M., and Nagel, S. R., 'Physics of the granular state', *Science* 255 (1992): 1523.

Jaeger, H. M., Nagel, S. R., and Behringer, R. P., 'The physics of granular materials', *Physics Today* (April 1996): 32.

Jullien, R., and Meakin, P., 'Three-dimensional model for particle-size segregation by shaking', *Physical Review Letters* 69 (1992): 640.

Kemp, M., *Visualizations* (Oxford: Oxford University Press, 2000).

Kerner, B. S., *The Physics of Traffic* (Berlin: Springer, 2004).

Kessler, M. A., and Werner, B. T., 'Self-organization of sorted patterns ground', *Science* 299 (2003): 380.

Knight, J. B., Jaeger, H. M., and Nagel, S. R., 'Vibration-induced size separation in granular media: the convection connection', *Physical Review Letters* 70 (1993): 3728.

Krantz, W. B., Gleason, K. J., and Caine, N., 'Patterned ground', *Scientific American* 259(6) (1988): 44.

Krause, J., and Ruxton, G. D., *Living in Groups* (Oxford: Oxford University Press, 2002).

Kroy, K., Sauermann, G., and Herrmann, H. J., 'Minimal model for sand dunes', *Physical Review Letters* 88 (2002): 054301.

Kroy, K., Sauermann, G., and Herrmann, H. J., 'Minimal model for aeolian sand dunes', *Physical Review E* 66 (2002): 031302.

Lancaster, N., *Geomorphology of Desert Dunes* (London: Routledge, 1995).

Landau, L. D., and Lifshitz, E. M., *Fluid Mechanics* (Oxford: Pergamon Press, 1959).

L'Vov, V., and Procaccia, I., 'Turbulence: a universal problem', *Physics World* 35 (August 1996).

Machetel, P., and Weber, P., 'Intermittent layered convection in a model mantle with an endothermic phase change at 670 km', *Nature* 350 (1991): 55.

Makse, H. A., Havlin, S., King, P. R., and Stanley, H. E., 'Spontaneous stratification in granular mixtures', *Nature* 386 (1997): 379.

Manneville, J.-B., and Olson, P., 'Convection in a rotating fluid sphere and banded structure of the Jovian atmosphere', *Icarus* 122 (1996): 242.

Marcus, P. S., 'Numerical simulation of Jupiter's Great Red Spot', *Nature* 331 (1988): 693.

Melo, F., Umbanhowar, P. B., and Swinney, H. L., 'Hexagons, kinks, and disorder in oscillated granular layers', *Physical Review Letters* 75 (1995): 3838.

Metcalfe, G., Shinbrot, T., McCarthy, J. J., and Ottino, J. M., 'Avalanche mixing of granular solids', *Nature* 374 (1995): 39.

Morris, S. W., Bodenschatz, E., Cannell, D. S., and Ahlers, G., 'Spiral defect chaos in large aspect ratio Rayleigh–Bénard convection', *Physical Review Letters* 71 (1993): 2026.

Mullin, T., 'Turbulent times for fluids', in Hall, N. (ed.), *Exploring Chaos. A Guide to the New Science of Disorder* (New York: W. W. Norton, 1991).

Nickling, W. G., 'Aeolian sediment transport and deposition', in Pye, K. (ed.), *Sediment Transport and Depositional Processes* (Oxford: Blackwell Scientific, 1994).

Ottino, J. M., 'Granular matter as a window into collective systems far from equilibrium, complexity, and scientific prematurity', *Chemical Engineering Science* 61 (2006): 4165.

Parrish, J. K., and Edelstein-Keshet, L., 'Complexity, pattern, and evolutionary trade-offs in animal aggregation', *Science* 284 (1999): 99.

Parteli, E. J. R., and Herrmann, H. J., 'Saltation transport on Mars', *Physical Review Letters* 98 (2007): 198001.

Perez, G. J., Tapang, G., Lim, M., and Saloma, C., 'Streaming, disruptive interference and power-law behavior in the exit dynamics of confined pedestrians', *Physica A* 312 (2002): 609.

Potts, W. K., 'The chorus-line hypothesis of manoeuvre coordination in avian flocks', *Nature* 309 (1984): 344.

Rappel, W.-J., Nicol, A., Sarkissian, A., Levine, H., and Loomis, W. F., 'Self-organized vortex state in two-dimensional *Dictyostelium* dynamics', *Physical Review Letters* 83 (1999): 1247.

Reynolds, C., 'Boids', article available at <http://www.red3d.com/cwr/boids/>.

Reynolds, C. W., 'Flocks, herds and schools: a distributed behavioral model', *Computer Graphics* 21(4) (1987): 25.

Ruelle, D., *Chance and Chaos* (London: Penguin, 1993).

Saloma, C., Perez, G. J., Tapang, G., Lim, M., and Palmes-Saloma, C., 'Self-organized queuing and scale-free behavior in real escape panic', *Proceedings of the National Academy of Sciences USA* 100 (2003): 11947.

Schwämmle, V., and Herrmann, H., 'Solitary wave behaviour of sand dunes', *Nature* 426 (2003): 619.

Scorer, R., and Verkaik, A., *Spacious Skies* (Newton Abbott: David & Charles, 1989).

Shinbrot, T., 'Competition between randomizing impacts and inelastic collisions in granular pattern formation', *Nature* 389 (1997): 574.

Sokolov, A., Aranson, I. S., Kessler, J. O., and Goldstein, R. E., 'Concentration dependence of the collective dynamics of swimming bacteria', *Physical Review Letters* 98 (2007): 158102.

Sommeria, J., Meyers, S. D., and Swinney, H. L., 'Laboratory simulation of Jupiter's Great Red Spot', *Nature* 331 (1988): 689.

Stewart, I., and Golubitsky, M., *Fearful Symmetry* (London: Penguin, 1993).

Strykowski, P. J., and Sreenivasan, K. R., 'On the formation and suppression of vortex "shedding" at low Reynolds numbers', *Journal of Fluid Mechanics* 218 (1990): 71.

Sumpter, D. J. T., 'The principles of collective animal behaviour', *Philosophical Transactions of the Royal Society B* 361 (2005): 5.

Szabó, B., Szöllösi, G. J., Gönci, B., Jurányi, Zs., Selmeczi, D., and Vicsek, T. 'Phase transition in the collective migration of tissue cells: experiment and model', *Physical Review E* 74 (2006): 061908.

Tackley, P. J., Stevenson, D. J., Glatzmaier, G. A., and Schubert, G., 'Effects of an endothermic phase transition at 670 km depth in a spherical model of convection in the Earth's mantle', *Nature* 361 (1993): 699.

Tackley, P. J., 'Layer cake on plum pudding?', *Nature Geoscience* 1 (2008): 157.

Thompson, D'A. W., *On Growth and Form* (New York: Dover, 1992).

Toner, J., and Tu, Y., 'Long-range order in a two-dimensional dynamical XY model: how birds fly together', *Physical Review Letters* 75 (1995): 4326.

Tritton, D. J., *Physical Fluid Dynamics* (Oxford: Oxford University Press, 1988).

Umbanhowar, P. M., Melo, F., and Swinney, H. L., 'Localized excitations in a vertically vibrated granular layer', *Nature* 382 (1996): 793.

Umbanhowar, P. M., Melo, F., and Swinney, H. L., 'Periodic, aperiodic, and transient patterns in vibrated granular layers', *Physica A* 249 (1998): 1.

Van Heijst, G. J. F., and Flór, J. B., 'Dipole formation and collisions in a stratified fluid', *Nature* 340 (1989): 212.

Vatistas, G. H., 'A note on liquid vortex sloshing and Kelvin's equilibria', *Journal of Fluid Mechanics* 217 (1990): 241.

Vatistas, G. H., Wang, J., and Lin, S., 'Experiments on waves induced in the hollow core of vortices', *Experiments in Fluids* 13 (1992): 377.

Velarde, G., and Normand, C., 'Convection', *Scientific American* 243(1) (1980): 92.

Vicsek, T., Czirók, A., Ben-Jacob, E., Cohen, I., and Shochet, O., 'Novel type of phase transition in a system of self-driven particles', *Physical Review Letters* 75 (1995): 1226.

Welland, M., *Sand: The Never-Ending Story* (Berkeley: University of California Press, 2008).

Werner, B. T., 'Eolian dunes: computer simulations and attractor interpretation', *Geology* 23 (1995): 1057.

Williams, J. C., and Shields, G., 'Segregation of granules in vibrated beds', *Powder Technology* 1 (1967): 134.

Wolf, D. E., 'Cellular automata for traffic simulations', *Physica A* 263 (1999): 438.

Worthington, A. M., *A Study of Splashes* (London: Longmans, Green & Co., 1908).

Zik, O., Levine, D., Lispon, S. G., Shtrikman, S., and Stavans, J., 'Rotationally induced segregation of granular materials', *Physical Review Letters* 73 (1994): 644.